DISCOVERING
PLANT LIFE

Cover picture: A dew covered bud of a garden nasturtium Tropaeolum majus *reveals the beauty and intricacy of a young flower.*

Opposite: This tropical flower, a member of the mallow family, provides a colorful resting place for a tiny spider.

DISCOVERING PLANT LIFE
by P. Francis Hunt

Published by

STONEHENGE

in association with

The American Museum of Natural History

The author
Peter Francis Hunt is Principal Lecturer in Ecology and Conservation, and Senior Course Tutor in Landscape Architecture in the School of Architecture and Landscape at the Thames Polytechnic, Dartford, England. He has been a research taxonomist in the Orchid Herbarium of the Royal Botanic Gardens, Kew, and was a member of the Royal Society Expedition to the Solomon Islands.

The consultants
Anthony Huxley has been involved with horticulture as a writer, photographer and consultant since 1949. He is the author of thirty books, the latest being an encyclopedia of gardening, and has co-authored three successful books on mountain flowers, flowers of the Mediterranean and the flowers of Greece and the Aegean.

The American Museum of Natural History
Stonehenge Press wishes to extend particular thanks to Dr. Thomas D. Nicholson, Director of the Museum, and Mr. David D. Ryus, Vice President, for their counsel and assistance in creating this volume.

Stonehenge Press Inc.:
Publisher: John Canova
Editor: Ezra Bowen
Deputy Editor: Carolyn Tasker

Sceptre Books Ltd.
Editorial Consultant: James Clark
Managing Editor: Barbara Horn

Created, designed and produced by
Sceptre Books Ltd, London.

© Sceptre Books Ltd. 1979, 1983
All rights reserved. No part of this book may be reproduced or utilized in any form or by any means, electronic or mechanical, including photocopying, recording or by any information storage or retrieval system, without permission in writing from Sceptre Books Ltd.

Library of Congress Card Number: 81-50150
Printed in U.S.A. by Rand McNally & Co.
First printing

ISBN 0-86706-014-X
ISBN 0-86706-065-4 (lib. bdg.)
ISBN 0-86706-034-4 (retail ed.)

Set in Monophoto Rockwell Light by
SX Composing Ltd, Rayleigh, Essex, England
Separation by Adroit Photo Litho Ltd., Birmingham, England

Contents

The Evolution of Plants	6
The Structure of Plants	8
Roots	10
Stems	12
Leaves	14
Flowers	16
Fruits and Seeds	18
Bulbs, Corms and Tubers	20
Photosynthesis	22
Respiration and Transpiration	24
Plant Reproduction	26
Pollination	28
Germination	30
Plant Growth and Development	32
Plant Behavior	34
Plant Diversity	36
Classification and Names	38
Bacteria and Cyanobacteria	40
Viruses and Phages	42
Algae I	44
Algae II	46
Fungi I	48
Fungi II	50
Lichens	52
Mosses and Liverworts	54
Ferns and their Relatives	56
Palms	58
Habitats	60
Wetlands	62
Tundra and Alpine Plants	64
Deciduous Forests	66
Conifer Forests and Gymnosperms	68
Mountains and Island	70
Grasslands	72
Deserts	74
Tropics	76
Plants and Mankind	78
Plants as Medicines and Drugs	80
New Plants from Old	82
Growing Plants	84
Growing Trees	86
Growing Shrubs and Climbers	88
Herbaceous Plants	90
Conservation and Threatened Species	92
Glossary	94
Index, Credits and Bibliography	96

The World of Plants

Brightly colored marigolds and pansies in window boxes, roses and geraniums in parks and gardens, trees in full leaf, and fruit ripening in orchards are all part of the summer scene. In autumn the flowers fade and now the trees are clothed in red and gold. The delicate silhouettes of their leafless limbs is one of the beauties of the winter landscape. In spring comes the renewal of life as the first green shoots poke through the cold soil. Understanding how and why these changes take place is just one part of the fascinating study of plant life, or botany.

There are millions of plants in the world that are too small to see. Microscopic bacteria and viruses live in the air, the soil and the water. Looking like tiny bats and balls when greatly magnified, these minute, primitive plants can be both useful and harmful to all living things. The microscopic algae such as the diatoms and dinoflagellates that live in the seas are an important food for fish and other marine life. Seen through an electron microscope, they are startling in their beauty. Other kinds of algae in various colors and in all sizes from the single cell to the giant seaweed also populate the waters, and some of them have come ashore.

There are nations of plants spread around the world. The hot, dry deserts support sparse populations of cacti and other succulents. The marshes can be identified from a distance by their armies of reeds and rushes, and even the mountains and the arctic lands are not without plant life. In the forests the trees dominate the shrubs and herbaceous plants that grow under their broad canopy, and in the grasslands the waving stalks of cereal plants symbolize how important plants are to all other forms of life.

Birds and insects, reptiles and mammals—including humans—are dependent on plants not only to supply food, but also to replenish oxygen. Plants are a source of medicines and fuels and the materials for clothes, buildings and hundreds of other items in use every day. The botanist seeks to understand the nature of plants as well as to protect and improve them. This book is a basic guide to botany for everyone who wants to know what botanists have discovered and how plants grow.

The Evolution of Plants

*Remains of plants that lived and died millions of years ago can be preserved as a thin layer of carbon or as impressions in rock. Fossilized leaves, far left, are similar to those of the maidenhair tree (*Gingko biloba*). The leaf shape imprinted in rock, left, is from an early flowering plant.*

What the earth was like before life began is hard to imagine. Because there is no remaining evidence, scientists can only theorize. None of the flowering plants, trees, insects and animals of the world today existed billions of years ago. But once the surface of the new earth started to cool down, life began in the shallow seas. Ultraviolet rays from the sun, together with lightning, caused water to combine with methane gases and ammonia to form proteins and nucleic acids. No one knows whether this chemical reaction occurred only once or a series of such reactions took place during a hundred million years. What we do know for certain—because there are fossils to prove it—is that the chemicals produced combined to form organisms, very simple forms of life, that could grow and reproduce. Over a period of millions of years these plant-like organisms developed the ability to take energy from the sun, turn it into food and store it. This process involved the production of oxygen. The oxygen entered the atmosphere and in time other organisms that could use it evolved.

Some of the oxygen in the air changed into a special form called ozone and settled in the upper layers of the atmosphere. Ozone acts as a screen against harmful ultraviolet rays and other cosmic radiation. Protected by the ozone layer, the simple organisms continued to grow and adapt to a changing world.

The gradual development, or evolution, of plants occurs by the very slow and prolonged accumulation of tiny but important changes in their ability to survive. For example, all of the offspring of any plant are similar,

Millions of years ago tree trunks in Yellowstone National Park, Wyoming, left, were turned to stone by deposits of mud and minerals.

but some are slightly stronger than others. The stronger plants live longer and produce more offspring, while the weaker ones are less successful and gradually become extinct. We do not always know how or why these changes take place. Before man developed the medical expertise that he has today, this also applied to mankind—the strongest individuals survived while the weakest did not.

The earliest organisms, which lived three and a half billion years ago, were similar to the cyanobacteria that still exist today. We know very little about other plants that might have lived during the next two and a half billion years because the period produced few fossils. We know that life diversified greatly after that time because fossils from plants that lived only 440 million years ago show structures and life cycles much more complicated and varied than the early forms.

By about 350 million years ago giant horsetails, clubmosses and fern-like plants were widespread on the earth's surface. We call the areas in which they lived "coal forests" because their fossilized remains are the coal we mine and burn today.

Well-preserved fossils found in several parts of the world show that the first flowering plants appeared some 135 million years ago. Today flowering plants are by far the most abundant, but most of the early coal-forest types have become extinct.

Flowerless plants such as algae, fungi, mosses, liverworts and ferns still exist and are important in helping to maintain the balance of nature. But they have not remained unchanged since they first appeared on earth those hundreds of millions of years ago. They are still evolving, and so too are the flowering plants and animals. The evolution of plants never stops—it is a continuous process.

The Structure of Plants

The bodies of all plants and animals are made of cells. No matter how large a plant or an animal is, most of its cells are very small and invisible to the naked eye. Nevertheless, these cells perform all of the plant's essential activities. Even the smallest cells can manufacture food, while the larger ones usually absorb and store it. Plants use this food to produce energy to grow and create new cells. Some simple plants, such as bacteria and green algae, are made of only one cell. Most other plants are made of many different types of cells joined together and, therefore, are large enough to be seen without visual aid.

Most plant cells have a multilayered wall made mainly of cellulose. The wall provides support for the cell and along with the cell membrane inside it prevents the watery living material from oozing away, and also prevents harmful organisms from getting in. The living matter within the wall is composed of cytoplasm

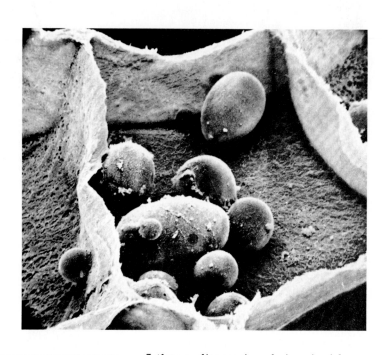

A three-dimensional view inside a dead cell of a potato plant, Solanum tuberosum, *is possible with a scanning electron microscope. The cytoplasm has disappeared revealing the cell walls and starch grains.*

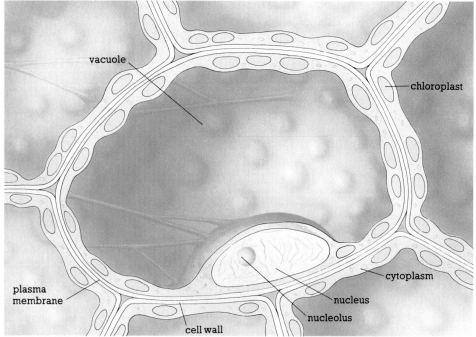

In the diagram at left is a typical cell of the green part of a plant with its major parts. The cell's main function is food manufacture by photosynthesis through the chlorophyll-containing chloroplasts.

The stigma sits on the top of the style of a flowering plant, in a magnified view.

In a greatly magnified view, a flower petal shows the details of its upper surface.

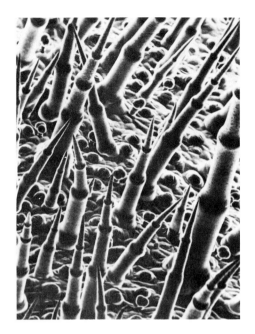

Spikes on the surface of a leaf appear in a magnification, but to the touch they have a downy feel.

and a nucleus. Cytoplasm is a mass of proteins, carbohydrates, water and inorganic chemicals—a chemical factory that enables the cell to live and work. The nucleus of the cell is the control center. It has coded instructions in the form of complex chemical compounds that control how fast the cell will grow and determine whether it will become part of a leaf, a flower, or a root. These compounds make up deoxyribonucleic acid, or DNA for short. The nucleus of a single cell contains about four miles of DNA arranged as a double coiled string of molecules. Exactly how its instructions are obeyed as the cells develop is still not fully understood.

When a cell reproduces, it divides into two equal parts, and each part must have its own nucleus in order to live. When a plant reproduces, the coded instructions in the nucleus make the new plant turn into a plant like its parents.

A simple, single-celled plant fulfills all the functions necessary to live, grow and produce new generations within its one cell. In more complex plants, many cells are joined together to form various tissues that perform different jobs. The "veins" of a plant, for example, are made of xylem and phloem cells joined end to end to form the tube-shaped vascular tissue. The xylem cells conduct water and minerals from the roots of the plant to the green parts where food is made. The phloem tissue transports the food to the rest of the plant including the roots. In certain plants, the cellulose in the walls of some xylem cells is partially replaced by strengthening layers of a chemical compound called lignin. The woody parts of the trunks, branches and twigs of trees and shrubs owe their strength and durability to the presence of large amounts of lignin.

Tissues too can join together to form organs, which have more complicated jobs to do. For example, the stamen of a flower is an organ that helps the plant to reproduce. But the stamen cannot do this job alone; it has to work with another organ. Two or more organs working together are called an organ system. A single flowering plant, whether it is a buttercup or a sunflower, is actually a complex team of organ systems made of millions of cells working together. In some plants, especially those growing in an extreme environment, such as a desert, some of the tissues and organs of plants may appear to be missing, but they are not. Either they persist for only a short time or they are reduced to a minute, unrecognizable form. For example, the spines on a cactus are actually highly modified leaves.

Roots

In most plants roots serve two major purposes. They anchor the plant and help it to stay upright, like the foundation of a building, and they provide it with water and minerals from the soil. Roots may have other functions as well. In some tropical climbing plants, especially those with short stems and very small, paper-like leaves, the roots of the plant rather than the leaves are responsible for the manufacture of food and are for this reason green. Roots can also be food storage organs, as in turnips and beets. Trees growing in tropical swamps may have roots that grow upwards out of the water and absorb oxygen. In some plants, such as English ivy, the roots help the plants to climb by clinging onto a suitable surface.

In a fibrous root system, such as that of the grasses, the base of the plant stem remains above the ground, rather than extending into the soil. Below the stem is a mass of small roots in the soil. In a tap root system the base of the plant stem goes below the ground and often becomes thickened. This can be clearly seen in dandelions and carrots.

Most plants have branching roots with a total length often exceeding six miles. Each root is not usually very long, although in certain tropical epiphytes aerial roots can be over ninety feet in length.

Inside all roots are the veins that carry the water and minerals to the stem, branches and leaves. In the roots the veins are concentrated into a compact core, like the lead in the center of a pencil. In the stems of some plants the veins are distributed around the edges like the wooden part of the pencil. Between the roots and the bottom of the stem is a transition zone in which the veins change over from their central position in the roots to the border position in the stem.

At the tip of a root is an actively growing mass of cells called the root meristem, which is protected by a layer of cells that form the root cap. The cells in the meristem grow rapidly, pushing the capped tip farther into the

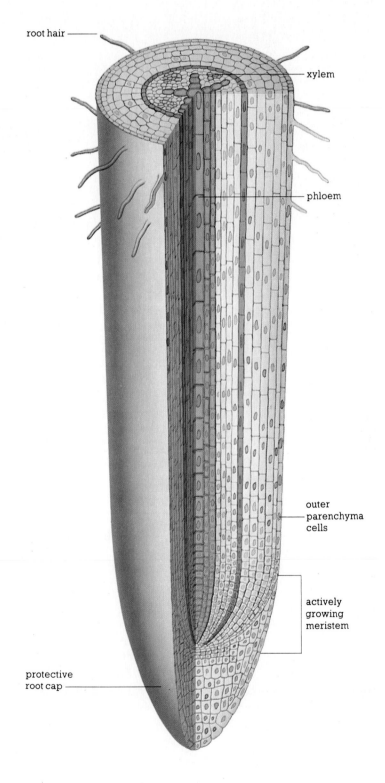

The cross section of a root magnified 100 times, right, shows the growing region, water-conductive veins, protective outer cell layer and root hairs.

Roots require air to carry out their functions, and most can find it in the soil. But in a waterlogged and airless swamp in South Carolina, left, a swamp cypress, Taxodium distichum, *has roots that protrude above the soil to take air from the atmosphere through special passages to the root system.*

soil to find fresh supplies of minerals and water and to gain greater anchorage. Further along the root, behind the cap and meristem is an area in which the outer cells become lengthened and extend into spaces between the particles of soil. These are the root hairs, the main water-absorbing parts of the roots. Root hairs are short-lived; they die as the root gets longer and new root hairs are formed. The plant thus has a useful system for supplying a constant source of food and water to its stems and leaves.

The withered outer cells of the rest of the root, and the layer just beneath them are corky, woody or waxy, so that when the soil is dry the root does not lose water and minerals. These tough layers also prevent pests from entering the roots.

The roots of many plants combine with threads of certain fungi to form a mutually beneficial association called a mycorrhiza. The fungus takes manufactured food from the plant and in return passes water and minerals from the soil to the roots. This interrelationship is called symbiosis. Trees, heathers, orchids and many other plants have the mycorrhizal fungi between the cells of the roots, and even inside the cells. Most orchids, in fact, need such fungi in order to grow.

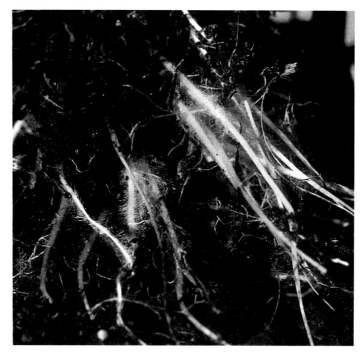

Shown in close-up above, the roots of plants penetrate the soil. Water is absorbed from the soil into the plant by the thin root hairs.

Stems

The stems of most plants usually stand upright, but how do they stay that way? Imagine many thousands of cells lying next to and on top of each other inside the stem. The walls of the cells are elastic. As the cells absorb water they expand and stretch until they are taut like balloons full of air. Now all the cells are pressing on each other and the stem wall. This pressure holds the stem up. When a plant does not get enough water, the cells shrink and the stem droops.

The stem of a plant supports its leaves and flowers; many stems also extend into the soil to support the root system as well. The stem contains the veins that transport minerals, water and food. Many plants have green stems that can also manufacture and store food, especially starch and oils. In some plants such as cacti, the stem stores water.

Stems are usually cylindrical, but some are square or even triangular. The surface of the stem can be flat or grooved, hairy or smooth.

The stem of a flowering plant often bears leaves. At points called nodes the stem produces a branch from which the leaves grow. Sometimes these branches have smaller branches of their own with a leaf growing

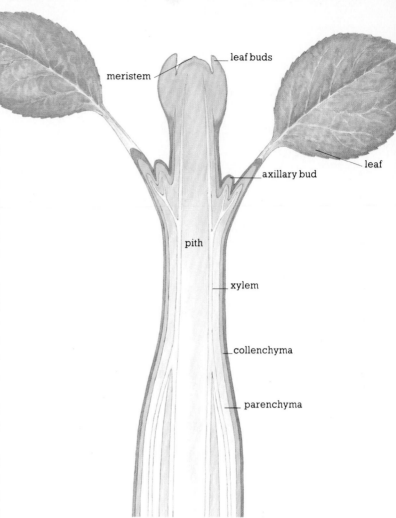

The "leaves" of butcher's broom, Ruscus aculeatus, below, are very unusual. They are actually flattened stems called cladophylls. The plant's true leaves are dried scales hidden by the flowers. In some species the leaf stalk, or petiole, is flattened like a leaf.

This diagram is a longitudinal section of a developing shoot. In addition to leaf-growing areas, it shows major internal tissues including the xylem, whose cells transport water and minerals upwards from the roots.

at the end of each. The dwarf shrub butcher's-broom has an unusual stem on which the leaves are reduced to minute scales, and the stem itself is flattened, leaf-shaped and green.

The stems on short-lived nonwoody plants generally die and decay along with the leaves, whereas on many shrubs and trees the stems continue to live when the leaves die. When the dead leaves fall off, they leave a scar on the branch.

At the top of each stem is a terminal growing point or bud. Other buds, called secondary buds, grow in the angles between the leaves and the stem. A secondary bud may produce a branch or a flower, or it may remain

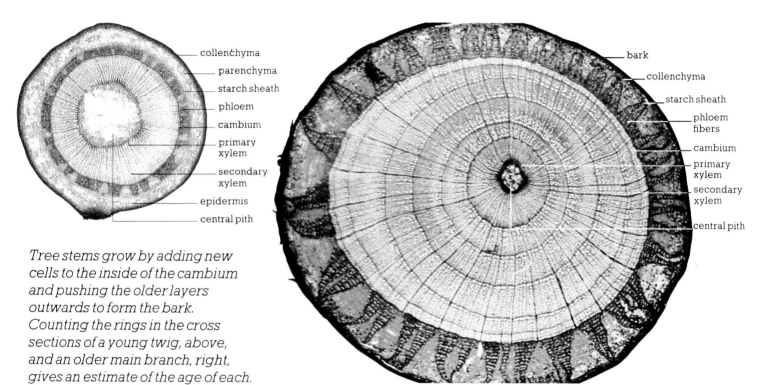

Tree stems grow by adding new cells to the inside of the cambium and pushing the older layers outwards to form the bark. Counting the rings in the cross sections of a young twig, above, and an older main branch, right, gives an estimate of the age of each.

dormant, or resting. Dormant buds either wither away or survive until the next growing season and grow into a branch or flower. If the terminal bud is damaged, the dormant bud can take its place as the main growing point of the plant.

All woody parts of trees increase in diameter as layers of new growth are added near both the center and the outside of the stem. The layer of cells that produce this new growth is called a cambium. In the cambium of the inner part of the stem chemical changes occur to form wood. This inner wood growth presses against the outer layers of the stem, which become strained and eventually crack. To replace this damaged layer another cambium layer, called the cork-cambium, produces an outer layer of elastic cork. In some species the cork-cambium produces cork layers that dry and are continually lost, thus leaving a smooth new surface on the tree. In many species, the cork dies but is not shed and progressively deeper layers of cork are added. In these cases the stem surface has a familiar scaly appearance. These layers of dead and living tissues are called bark. The bark helps woody plants to retain water and also protects them from animals, insects and fungi.

Many plants have climbing and scrambling stems. The stems of runner beans, for example, climb by coiling around a support, which is sometimes another plant. Blackberries scramble by hooking their sharp thorns into a wall or similar support, and vines and cucumbers have twining stem tips called tendrils.

Although most stems are above ground and grow towards the light, some, such as the tubers of potatoes, grow into the soil and become swollen with stored food. Certain plants, among them strawberries, have stems called runners that grow parallel with the soil surface and are used to reproduce the plant. Still others, such as irises, have stems that grow just below the soil surface and are called rhizomes.

Leaves

The leaf blade, or lamina, of a plant is connected to the stem by the leaf stalk, or petiole. The petiole continues into the leaf blade as the main rib, where it often branches into a network of veins. Where the petiole joins the stem is a small secondary bud, and below this there is often a small leaf-like outgrowth called a stipule. In some plants the stipule is large and completely surrounds the stem; in others it is modified to form a clinging tendril.

A leaf is constructed from layers of cells, and each layer has its own function. Leaves that grow in dark, damp rain forests may be only two or three layers thick. The whole outer layer of the leaf is called the epidermis, which, like a skin, mainly holds the leaf together. On both the top and bottom surfaces the epidermis generally has a waxy, waterproof outer film called the cuticle. The bottom surface also has small pores, called stomata, through which air and water pass. Inside the epidermis of the top of the leaf is a layer of cells filled with chlorophyll. This is called the palisade parenchyma, and it is the main food-manufacturing part of the leaf. Toward the lower surface of the leaf is a mass of irregularly shaped, spongy cells, called spongy parenchyma, which also can manufacture food and store it. The cells in these two layers are connected to the veins, and between them are many air spaces.

The manufacturing of food, or photosynthesis, takes place mainly in the leaves. In some plants leaves can store large quantities of food, enabling the plant to survive an unfavorable season. The leaves of most ferns and related plants bear the organs necessary for reproduction. The leaves of deciduous trees are shed

The bright colors of trees in the fall are caused by tannins and other substances lingering in dying leaves. Water-lily leaves float on the surface of a pond, and sacred lotus leaves stand above it.

Leaves are described according to their shapes, edges, or bases. In the picture at right, the veins of the leaves can be clearly seen. In grass, second from left, the veins run parallel along the leaf. In all the other plants there is a main central vein with side branches.

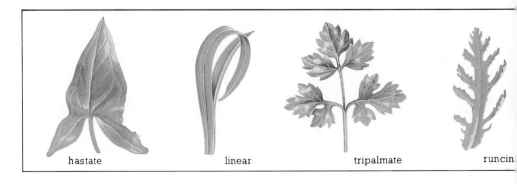

hastate linear tripalmate runcin

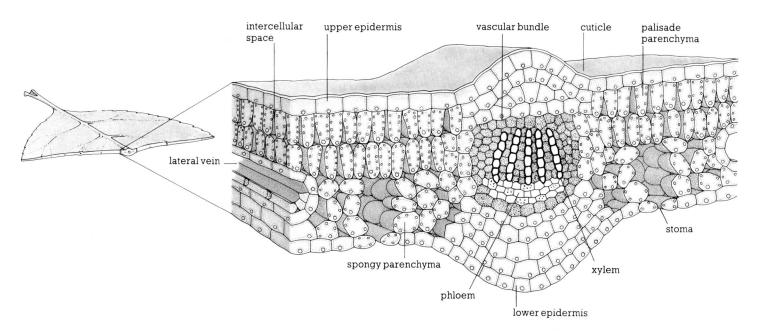

in the autumn, as photosynthesis stops. Just before they fall they usually change color. The food reserves of these leaves are mostly absorbed by the tree, and materials remaining in the leaves, such as various pigments, can be seen. So, in temperate regions, just before the trees lose their leaves the forests are aglow with red- and yellow-leaved trees.

Leaves vary greatly in size. In some water plants they are as small as this letter o, while in certain tropical land plants they may be as long as sixty-five feet. The shapes, edges, tips and bases of leaves also vary greatly. Some of these variations seem to serve no purpose, but they may help prevent the plant from losing water or help to expose as much of the leaf as possible to sunlight.

When a plant absorbs too much water from the soil, the excess is sometimes given off from the edges of the leaves. In droughts the leaves of many plants store

Leaves vary in shape and size but not in basic structure. Above, the midvein branches to each part of the leaf, carrying water and nutrients to the cells and taking away the food made by photosynthesis. A cross section through the center of the leaf shows the conducting cells of the midvein, the outer protective layers and the photosynthetic cells in the two parenchyma layers.

water and become thick and succulent. Some leaves have become modified into long, thin, entwining tendrils, such as those of sweet peas. A most unusual modification are the leaves of the various pitcher plants, in which the leaf forms a container with liquid at the bottom. Insects are attracted by nectar near the top of the leaf. Once inside it, the insects are trapped and drown. Their bodies are then digested and absorbed into the plant.

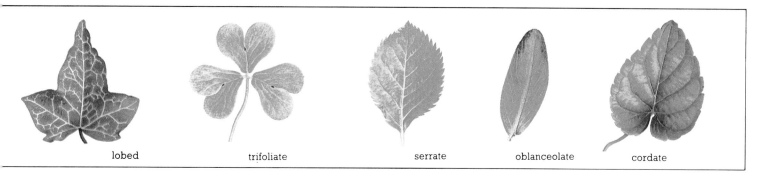

lobed   trifoliate   serrate   oblanceolate   cordate

Flowers

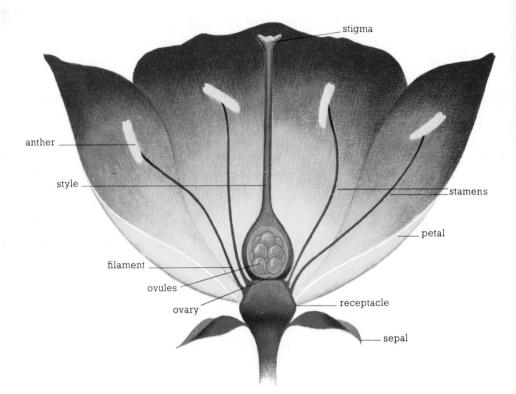

An unusual X-ray of a spray of roses shows how flower petals protect the reproduction area.

A flower with both male and female reproductive organs, in cross section, left, shows the female pistil surrounded by male stamens. The petals of the flower protect the reproductive organs and attract pollinating insects.

Flowers are the reproductive structures of certain seed-producing plants called angiosperms. Flowers have male and female cells, and when these unite, seeds are produced. The angiosperms are not the only seed-producing plants. Another group, the gymnosperms, produce seeds, but with cones rather than flowers. Flowers have a male organ called a stamen and a female organ called a pistil. The male stalk is called a filament, the female stalk is called a style. The male and female stalks are usually found together on a flower, but in some species they grow on separate flowers on the same plant, and in others they grow on separate plants.

The stamen bears the male cells in pollen in the anther at its tip. The female cells, called ovules, are contained in an ovary at the base of the pistil. Connected to the ovary by a slender stalk called a style is the sticky stigma used to catch the pollen.

The young, unopened flower is normally protected by a ring of special leaves called sepals. As the flower opens, the petals – another kind of leaf – are revealed. Their function is to protect the male and female organs and to help pollination by attracting animals. The petals and sepals can be separate or joined, different from each other or exactly the same. Usually, all the petals of a flower are similar, but in orchids, for example, the middle petal is quite different from the others.

Some species rely on the wind for pollination. Grasses and trees in temperate regions are usually wind-pollinated and may have separate male and female flowers, often on separate twigs. The male flowers are often larger than the females and release large quantities of fine, powdery pollen in the lightest of breezes. The small female flowers have feather-like stigmas that catch pollen. Wind-pollinated plants include birches, hazels, alders and most of the conifers. Some aquatic plants are water-pollinated. The pollen is released into the water and carried by the current until

A rose, far left, is commonly pink in color. Geraniums, left, come in a range of pinks and reds. Yellow and white are the most usual colors of flowers, but red, pink blue and mauve flowers are also found in most plant families. No truly black flower has been found, and no gardener has yet been able to breed one.

trapped by the long, trailing underwater stigmas of the female flowers. Most species of flowering plants, however, are pollinated by insects such as ants, bees, butterflies, and beetles. Other invertebrates, such as spiders or even snails, can act as pollinators. Some of the more brightly colored flowers of the tropics are pollinated by small birds, particularly hummingbirds and bee eaters. Bats pollinate many night-flowering plants, and in the Amazonian forests of South America frogs are the pollinating agents.

Usually the shapes, colors and markings of the petals and sepals of a flower attract pollinators, but movement and scent can also be important. Often two or more attracting features act together. Many flowers secrete a sugary liquid called nectar at the base of the petals or sepals and occasionally elsewhere. Insects and birds visit the flower to feed on the nectar and the pollen that becomes attached to their bodies is transferred to the next flower they visit. Thus, a number of flowers can be pollinated in one day.

Some plants have single flowers on a stem, while others bear their flowers in a group, called an inflorescence. The main purpose of these groups seems to be to attract more insects than a single flower can.

A field filled with wild flowers, left, is probably one of nature's most beautiful sights. Classified as the angiosperms, flowering plants are the result of millions of years of plant evolution.

Fruits and Seeds

The function of flowers is to produce seeds that will ensure that the species continues from generation to generation. Many plants produce a great number of seeds to ensure the continuation of the species. The seed of most plants consists of an undeveloped plant, or embryo, with one or two tiny seed leaves and minute roots. Food is stored either in the seed leaves or in the tissue surrounding the embryo. The embryo and tissue are protected by the seed coat or testa. Some seed coats are porous and allow water to reach the embryo when it starts to grow.

The seeds of the orchid have a tiny embryo and no supply of food. To make sure that the species survives, a single orchid plant produces millions of seeds in one season. These seeds are so small that sixty million of them weigh less than a postcard. The seeds of most other plants, however, are larger. Those of the coconut tree, for example, can be as big and as heavy as a bowling ball.

Some seeds are protected only by the testa until they begin to grow, but most seeds are enclosed in fruit. The fruit can be dry, such as ripe pea pods and bean pods, or fleshy, such as apples, pears, plums, tomatoes, oranges, lemons and grapefruit – all foods for people. The seeds develop when the cells in the ovary and surrounding tissues take in moisture and begin to grow. The cells produce sugars and chemicals that give the fruits the flavors and colors that attract the animals that eat them, and in this way help to disperse the seeds.

Seeds must be scattered as far as possible from the parent plant so they do not compete with it for water, minerals and space. Plants have many devices for making sure that this happens. Some seeds and fruits, like those of the chestnut tree, have hooks that catch on

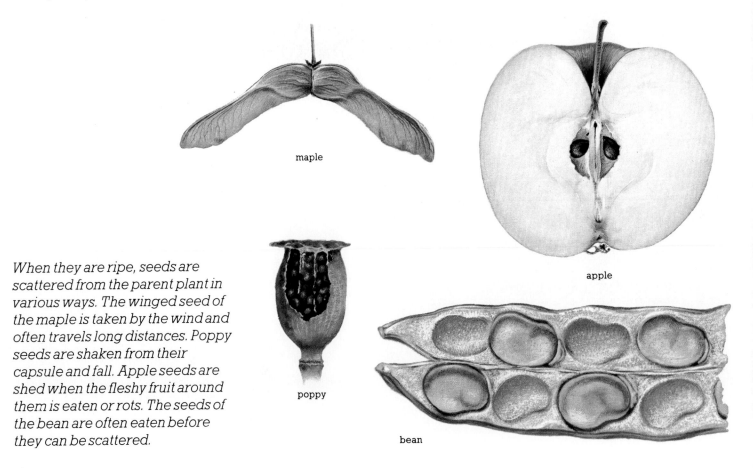

maple

apple

poppy

bean

When they are ripe, seeds are scattered from the parent plant in various ways. The winged seed of the maple is taken by the wind and often travels long distances. Poppy seeds are shaken from their capsule and fall. Apple seeds are shed when the fleshy fruit around them is eaten or rots. The seeds of the bean are often eaten before they can be scattered.

A gust of wind blows the seeds from a dandelion, left. With the aid of their feathery parachute, the seeds can travel a considerable distance. Dandelion seeds are plentiful because they can be made without the complications of sexual reproduction.

Apples, plums, peaches, cherries and loganberries, below left, are edible fruits enclosing seeds. Hard, prickly coats surround the seeds of the horse chestnut tree, below.

the fur and feathers of animals and birds or on people's clothing and are taken wherever the carrier goes. Sooner or later they fall or are brushed off and often land on the ground and begin to germinate. Some seeds are carried great distances on a bird's foot or a person's shoe or even among the dust and dirt of trucks, ships and aircraft. Seeds can also be transported inside an animal that has eaten such fruits as grapes, plums and pears. The seeds are released when the animal excretes its waste material.

The seeds of water plants, such as the floating water lilies, contain light corky or oily substances so that they can float and be scattered by the water.

Some seeds are transported by the wind. Very small seeds are simply blown around by air currents. Larger ones, like those of the dandelion, have feathery parachutes or, like those of sycamore trees, wings to help them scatter in the wind. Some plants, such as the poppy, shed their seeds when the wind shakes the dried capsule that holds them. Other capsules explode when the part of the fruit carrying the seeds dries and the pressure inside becomes too great. The seeds are shot into the air as if by a catapult and sometimes carried over great distances by the wind.

Bulbs, Corms, and Tubers

Many plants have developed organs that permit them to survive underground during an unfavorable season, such as a cold, dry winter or a hot, dry summer. These organs store food and water and also reproduce the plant. There are three main kinds of such organs – bulbs, corms and tubers. Bulbs are storage and reproductive organs that are formed underground at the end of the growing season and stay in the dormant, or resting, state until good growing conditions reoccur. The stems of bulbous plants, such as onions and shallots, daffodils and hyacinths, grow long only at flowering time and die quickly when the plant has finished flowering and fruiting. Meanwhile, one of the buds of the plant grows into a short stem. At the end of the growing season the bases of the leaves surrounding it become swollen with food reserves. The plant thus dies except for the short stem and the swollen, fleshy leaves of the bulb.

When certain plants, such as the gladiolus and the crocus, have flowered, the bases of their stems become swollen with food reserves and form corms. The leaves around these stems die, leaving papery scales on the outside of the corm. Corms store food not in swollen leaves like bulbs, but in the swollen stem. When conditions are suitable for the plant to bloom again, a bud at the base of the previous year's dead main stem starts to grow and eventually produces leaves and flowers.

Tubers are either swollen roots, as in the dahlia, or swollen stems, as in the potato. Some tuberous plants store food in their stems or roots only at the end of the growing season, but others continually use these

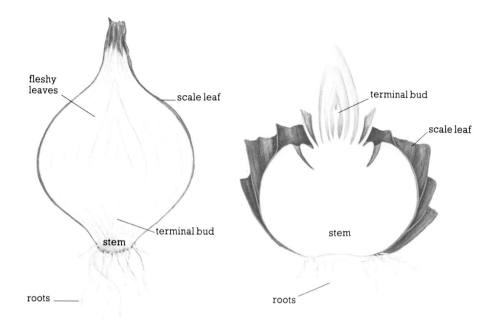

The stem of a bulb, far left, is reduced to a disk from which arise fat, fleshy, colorless leaves. In this state, below the ground and protected by an outer layer of dried, brown scaly leaves, bulbs can survive hot, dry summers and damp, cold winters.

A corm, left, differs from a bulb in that the condensed stem is much larger than the leaves. Axillary buds will either form small flowerless stems or may develop into small corms.

organs for storage. Buds from which shoots can grow appear on the outside of tubers. When most tubers are broken into smaller pieces, any piece with a bud will develop into a new plant.

People have used bulbs, corms and tubers, such as onions, taros and potatoes, for their food supply for thousands of years. Many others, such as ginger and licorice, are used as drugs or flavorings for foods. The foods stored in these organs are usually starches and sugars, and are often accompanied by other chemicals. The edible bulbs, corms and tubers are eaten immediately or kept for future use. They can also be transported great distances and planted far from their countries of origin.

Many decorative varieties of bulbs, corms and tubers, such as lilies, tulips, irises and gladioli, are grown commercially and are important exports. In many countries large areas are devoted entirely to producing them for sale around the world, and new varieties are bred each year. Some of them sell for high prices when they are first introduced to the general market. Not long after tulips were introduced into Europe from Turkey at the end of the sixteenth century, they became so popular that a craze called "tulip mania" swept Holland and neighboring countries.

The tubers of some plants are oddly-shaped and even look like parts of the human body or certain animals. Some people believe that such plants and their tubers and extracts can be used to cure diseases that affect the part of the body they resemble.

Potatoes are stem tubers. They were introduced to Europe in the sixteenth century from Peru and subsequently bred to create hundreds of different cultivated forms.

The Darwin type of garden variety of tulip shown here is very popular. Many thousands of tulip varieties are grown and there has been a considerable international trade in tulip bulbs since the sixteenth century.

Photosynthesis

Photosynthesis is the process by which green plants use energy from sunlight and carbon dioxide from the air to make food. Every green leaf, no matter how tiny, performs the miracle of photosynthesis. It makes all life on earth possible. A few specialized, minute plants and animals manufacture food differently, using simple substances. All animals need green plants for food; even meat-eating animals get food by eating animals that have eaten plants. So green plants are the primary producers of all the food in the world. Even though scientists can build machines to support life in space, no one has been able to duplicate the process of photosynthesis in a green leaf.

The substance that makes plants green is called chlorophyll. Molecules of chlorophyll are contained in parts of the plant cells called chloroplasts. When light, or radiant energy, falls on the green areas of a plant, the chlorophyll molecules trap it and generate minute electric currents. The plant uses this electricity to split the water molecules in the plant into oxygen and hydrogen atoms. The hydrogen atoms combine with carbon dioxide, which the plant has absorbed from the air, to make sugar, or glucose. The oxygen atoms are sent back into the air through the leaf's stomata. This is a simplified description of photosynthesis. In fact it is a much more complex biochemical process, in which many other substances are produced during intermediate stages. Scientists are trying to breed plants that produce large quantities of these intermediate substances, which could be of great value as industrial

The cells of green plants capture, store and change energy far more efficiently than any of man's most sophisticated technologies. In photosynthesis, diagrammed at right, energy from the sun enters a plant's cells and is used by the chlorophyll in the cells to change water and carbon dioxide from the air into energy-containing sugars. In the process, oxygen is produced and released into the air. The sugars are then passed into the stems and carried to other parts of the plant, where they are stored or used for growth and repair. The process of photosynthesis takes place continuously in all green plants.

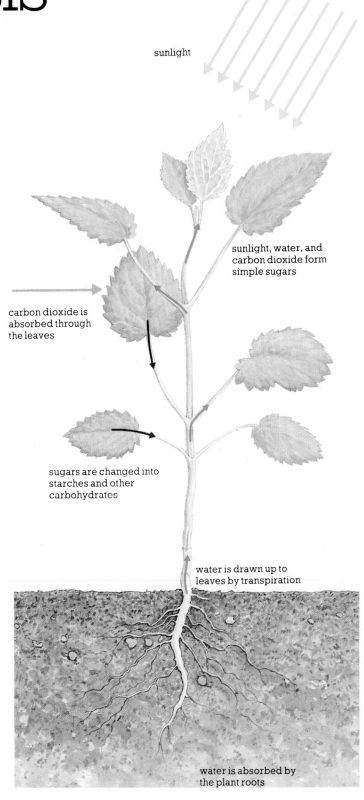

sunlight

sunlight, water, and carbon dioxide form simple sugars

carbon dioxide is absorbed through the leaves

sugars are changed into starches and other carbohydrates

water is drawn up to leaves by transpiration

water is absorbed by the plant roots

The edible mushroom Morchella esculenta, *or morel, far left, is a nongreen plant that cannot make its own food by photosynthesis but obtains it from decaying plants. Some plants, such as variegated ivies and* Fittonia argyroneura, *left, manufacture food by photosynthesis in their green parts and store it in their nongreen areas.*

raw materials and fuels in the near future.

The light energy used for photosynthesis stays locked in the sugars and starches as chemical energy until it is needed to help the plant move, grow and reproduce, or until the plant is eaten or dies. The glucose formed as the end product of photosynthesis can be converted by the plant into other organic compounds necessary for life, such as cellulose, lignin, oils and proteins. Bacteria, fungi and microscopic animals living in soil feed on dead plants and animals. The waste products of these animals include carbon dioxide and water, which are then used by green plants for photosynthesis.

If the remains of dead plants fall into acid water or are covered by sand and mud, they may be preserved, and after many thousands of years they become fossils. Coal, oil and peat are made up of fossilized plants. When we burn them, we release stored energy.

Trees and shrubs have always been grown and collected by man as fuel for domestic heating and cooking and for industry. Today, nonwoody plants are being grown experimentally to assess their value as fuels of the future. In Hungary a new electrical generating station is to be fueled exclusively by the common reed,

In ponds, green blanketweed algae such as Cladophora *species, right, are photosynthetic plants. They use the abundant water and carbon dioxide and produce oxygen as a waste produce, which appears on the surface as bubbles.*

Phragmites australis, which is cultivated in mud flats along the Danube.

By using carbon dioxide and producing oxygen in photosynthesis, green plants help to maintain the balance between these two elements in the earth's atmosphere. If too many green plants are destroyed, the earth's atmosphere could become unbalanced. However, there is no need for immediate concern because the latest research indicates that life will continue for about another five billion years. By then forms of life that utilize other substances for fuel and energy may have taken the place of the green plants.

Respiration and Transpiration

Respiration is the process by which a plant absorbs oxygen from the air and releases the energy that was locked in sugars through photosynthesis. The plant uses the energy to fuel its growth and behavior. Respiration takes place in almost all plants and animals. Burning fossilized plants, such as coal or oil peat, is a process of rapid respiration, in which large quantities of oxygen are absorbed and a lot of energy is given off in the form of heat. Respiration in plants, then, is the burning of its sugars.

Respiration takes place in the cytoplasm of the plant's cells. Each molecule of sugar, or glucose, is broken down in a series of steps to make pyruvic acid, as energy is released. Still more energy is released as the acid undergoes further chemical changes until finally carbon dioxide and water are formed.

Transpiration is the process by which plants give off water through their leaves. Water is vital to plants for their food-making process and also to keep the stems and leaves rigid and the plant cool. The plant gathers water from the soil through its roots. The water passes up the stem through the xylem and to the branches and the leaves. Tremendous pressure is needed to raise water to the top leaves of the tallest trees. A plant can raise water to the height it requires because of the pressure created by the loss of water, through transpiration, in the leaves.

Water is lost from the plant's leaves chiefly by the action of guard cells on either side of the stomata, usually found on the lower epidermis. When the sun is shining or the air is damp, the guard cells absorb water vapor from other cells in the leaf. The water changes the shape of the guard cells and forces the pore (stoma) to open to allow water vapor to escape from the leaf. In the dark or when the air is dry, the guard cells may become limp and the pores close. Some plants also have a biological clock that opens and closes the stomata regardless of temperature, light or humidity. Stomata also facilitate the flow of gases, such as oxygen and carbon dioxide, used in the processes of respiration and photosynthesis.

In high humidity the water in a saturated plant cannot evaporate easily into the air. The plant copes by exuding drops of water from special cells at the edges of leaves. This process can often be seen on garden plants when the air is very damp on a cool morning.

A parsley leaf, right, and a carrot leaf, far right, have been magnified hundreds of times to show their pores, or stomata, in both the open and the closed positions. Carbon dioxide enters the plants through these stomata, and oxygen is expelled through them as a waste product. Water is also lost through the stomata in a process called transpiration. Plants depend on water for their rigidity and the transport of food materials.

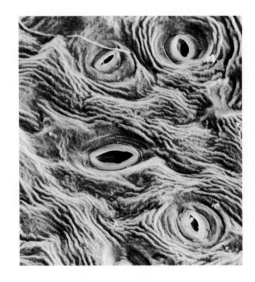
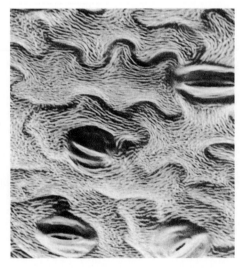

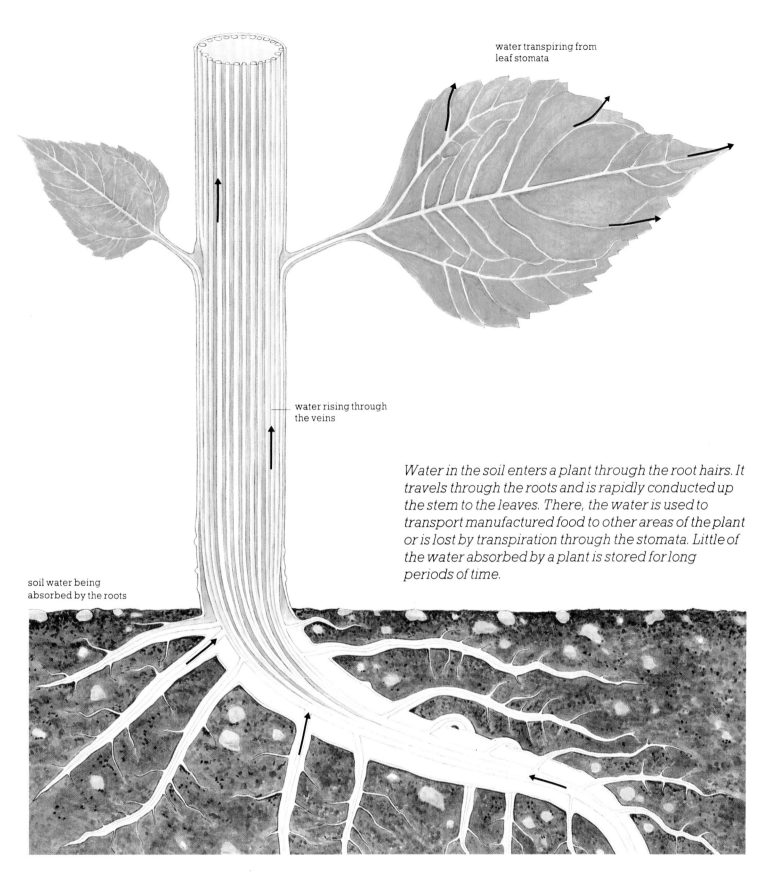

water transpiring from leaf stomata

water rising through the veins

soil water being absorbed by the roots

Water in the soil enters a plant through the root hairs. It travels through the roots and is rapidly conducted up the stem to the leaves. There, the water is used to transport manufactured food to other areas of the plant or is lost by transpiration through the stomata. Little of the water absorbed by a plant is stored for long periods of time.

Plant Reproduction

Plants reproduce either sexually or asexually, and some plants reproduce in both ways. Sexual reproduction requires separate male and female plants, or just male and female parts of a single plant, which produce sex cells called gametes. The gametes must come together, fuse and then grow in order to produce offspring. A much simpler form of sexual reproduction can be seen in certain thread-like algae. Two identical algae line up next to each other and, where they touch, their cell walls break down and the contents of one cell enter the other to produce offspring. Other algae, such as the brown seaweeds, develop special chambers that release male or female cells into the sea where they drift on currents and, perhaps, meet and unite.

Mosses and liverworts produce spores that develop into miniature plants. These in turn produce male and female gametes that eventually meet and fuse to make a new spore-producing plant. This process – a plant that produces spores and a plant that produces gametes – is called the alternation of generations.

Ferns reproduce in a similar way. Spores contained in protuberances on the underside of the leaves are shed and then germinate into minute, flat green plants.

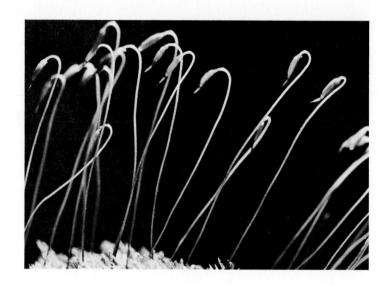

Mosses reproduce by releasing spores from capsules at the tops of long stems, above. The spores germinate to produce new moss plants.

This series of microphotographs shows a mature spore capsule of a fern leaf splitting open and casting the spores into the air.

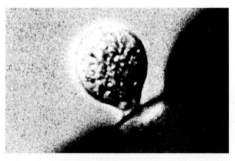  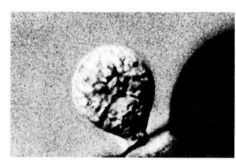

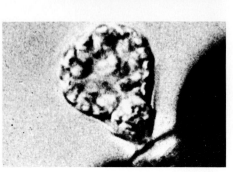  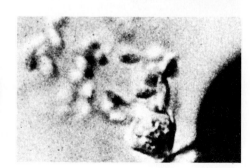

Many plants, such as this Kalanchoe tubiflora, *reproduce asexually, which means that they produce offspring from the existing plant without sexual reproduction. The spider plant is another common example.*

These plants produce gametes, which unite to form a new spore-bearing fern.

Reproduction by spores is a wasteful and uncertain method. A spore can retain its ability to grow as long as forty years after it has been released from the plant, but when it starts to grow it can survive only if conditions are favorable because it has no food reserves.

About 350 million years ago the first plants bearing seeds, rather than spores, appeared on earth. These early plants were the seed-ferns, and although they existed for over 50 million years, they eventually became extinct. Their place was taken by the naked seed plants, or gymnosperms. About 135 million years ago, the closed seed or flowering plants – technically, angiosperms – evolved.

Flowering plants in the wild usually reproduce sexually, but especially in extreme and harsh conditions, such as mountain tops, they may reproduce asexually. In such environments, sexual methods – which often depend on insects and birds to bring the gametes together – are unreliable. Asexual reproduction is accomplished by a part of a plant detaching itself from the parent to form a new plant.

Gardeners, nurserymen and agricultural scientists have greatly developed these asexual, or vegetative, methods of plant reproduction so that they can quickly produce large numbers of identical plantlets of a single kind. Today it is simple to produce millions of young plants, such as orchids, in a few weeks from a single parent plant. This is done by continually splitting and developing the growing point, or meristem, under laboratory conditions so that it develops into a miniature plant ready for sale. If smaller numbers of plants are required, the old techniques of taking cuttings or splitting the rootstock are used.

There are many advantages with cultivating and harvesting a crop of identical plants, but sometimes a disease or abrupt climatic change can do much damage because every plant is affected equally. In nature it is advantageous that plants reproduce sexually because the dominant characteristics of each parent can be combined in the offspring. Each species then exists in a range of forms, which greatly reduces the chance that any pest, disease or sudden change in climate is likely to be uniform in its effect. In this way the robust plats are able to carry on the species.

Pollination

In order for a flower to produce a new plant, the pollen from the stamen, the male stalk, must be transferred to the sticky stigma of the female pistil. This movement of pollen is called pollination. It is followed by fertilization – the male cells carried in the pollen and the female cells in the ovary unite to form a seed. Fertilization can take place only if the male and female cells are from the same or related species. The ovary must be at just the right stage of growth. A grain of pollen sends a tube containing male sex cells through the stigma and the style into the ovary. As soon as the sex cells unite, the seeds begin to develop.

The tiny grains of pollen can be carried to the stigma by the wind, water or an animal – even a snail, frog or bird – but they are usually carried by an insect. Chances are slim that a grain of pollen blown through the air or carried by a current of water will reach a stigma at the right time. Therefore plants pollinated by wind and water usually produce a great deal of pollen, so that at least some of it will find its target. Birch, poplar and hazel trees are wind-pollinated plants. Their many dangling catkins shed their pollen as they are shaken by the wind. Most grasses are wind-pollinated, too. The large amounts of pollen they produce cause hay fever, an allergic reaction that produces sneezing, runny nose and sore, itchy eyes. In the water plants that pollinate underwater, the pollen is carried by currents caught by long, trailing stigmas.

Plants that depend on animals to help them pollinate do not have to produce as much pollen as those that depend on wind or water. These plants have developed many features to attract animals and insects, among

Attracted by the shapes, colors or scents of flowers, insects visit individual blossoms in search of nectar or pollen as food. This picture shows how pollen clings to the insect's body – in this case a bee – so that when the next flower is visited, some of the pollen is transferred to the flower. The transfer of pollen by insects is the most common means of cross-pollination.

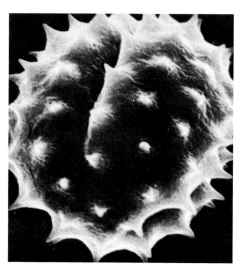

Pollen grains, top left, look just like specks of dust when they are shed from a catkin. However, when they are magnified, as seen in the other three pictures, the pollen grains of different plants show distinct features. Even when the living portion of the grain has decayed, the grain often retains its shape. By identifying and counting numbers of pollen grains at different depths in peat deposits, it is possible for botanists to identify the plants and vegetation that grew many thousands of years ago. Pollen stratigraphy, as the technique is called, is a major research tool in many fields, including archaeology.

them the colors and patterns of markings on their petals. Some plants give a reward of food to the animals that help them pollinate. This food can be nectar – a sweet, sugary liquid made inside the flower – or the pollen itself. As the insects gather their food, some of the pollen sticks to their bodies and is brushed off on the stigmas as the insects move from flower to flower looking for more food.

Some flowers attract insects through scent. The flower can smell pleasant and attract butterflies and moths, or it can smell like rotting animal flesh, and attract blowflies or beetles.

In some plants the insect-like shape of the flower attracts male insects, which try unsuccessfully to mate with the flower and in the process become covered with pollen. Other flowers have long, waving tassels that catch flying pollen. Many plants have a combination of all these features. Although insects are the most common pollinators, spiders, slugs and snails, frogs, bats and hummingbirds also pollinate certain plants.

Pollination does not always occur between two flowers. Many plants self-pollinate; that is, the pollen is transferred from the stamen to a stigma within the same flower. An insect or the wind can cause self-pollination, but it typically occurs when ripe stamens simply drop pollen onto a receptive stigma. Since less healthy plants can result from self-pollination, many plants have mechanisms that inhibit it. In some, the pollen is liberated a few days before the stigma becomes receptive. In others, a genetic mechanism prevents a pollen grain that has landed on a stigma in the same flower from sending a pollen tube into the ovary.

Germination

When the air, temperature, amount of moisture and light are suitable, the embryo plant in a seed can grow to become a mature plant. The early stage of growth is known as germination. It can begin immediately after a seed or spore has been shed by the parent, but most seeds and some spores can stay dormant for a time.

As a dormant seed usually has little stored water, its rate of respiration is low, and it is tolerant of extremes of temperature. The first step in germination is for the seed to absorb water and oxygen so that respiration can speed up. The cells in the embryo then start to expand and soon begin to divide. They use the food stored in the seed as their source of energy. As soon as the embryo has grown too large and the seed has taken in enough water, the testa splits and the young root and shoot emerge.

Warmth is usually required for germination. However, the seeds of some plants need a short period of freezing temperatures for the seed coat to split and germination to begin. The seeds of certain plants of dry, hot areas or within habitats subject to natural burning germinate more readily if they have been stimulated by high temperatures or fire. Some seeds germinate only in the dark, others require light, while still others germinate in the dark or the light. Seeds with limited amounts of stored food, such as those of orchids, will germinate only in the presence of a fungus that provides the necessary nutrients to sustain the developing embryos. In cultivation these seeds have to be germinated on a special jelly, with the correct nutrients added and sterilized to prevent unwanted fungal and bacterial growth. Germination of these tiny seeds can be a lengthy process, often taking up to a year for a shoot to appear. In most flowering plants and conifers, however, germination is accomplished much quicker, about ten days being the average.

Root hairs develop so that the young plant can take water and minerals from the soil. The seed leaves, called cotyledons, can remain underground, but in many plants they are pushed up on the growing shoot. They turn green as chlorophyll develops, at which point they begin to make food by photosynthesis.

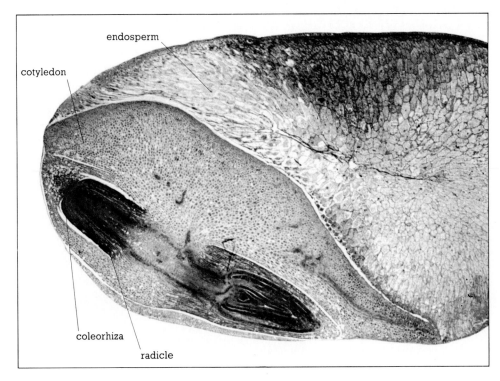

In this cross section of a seed of corn that is just beginning to germinate, the large cells of stored food look like scales on the top layer – the endosperm. The developing root, or radicle, is the rounded mass at the left, and the developing shoot is the pointed mass on the right.

A number of Acacia seeds have started to germinate. In some the seed leaves are just emerging, while in others the stems have started to stand up. In the center is one plantlet with an upright stem and two seed leaves clearly visible. Next to it is an older plantlet with a strong, straight stem that has produced its first true leaves.

Eventually the food stored in the cotyledons is used up and they wither away. By this time the first true leaves have grown and take over the manufacture of food.

The flowering plants are divided into two groups according to the number of cotyledons. The monocotyledons have a single cotyledon, and they are the grasses, palms, lilies and orchids. The dicotyledons, the larger group, have two. Most trees and shrubs and also flowers such as daisies and buttercups are dicotyledons. The monocotyledons and dicotyledons also differ in their flowers, leaves and internal structure. The growing points on the shoot and the root continue the growth of the plant above and below the ground until all traces of the seed have disappeared.

Some seeds can be eaten when they are dormant such as corn, wheat and rice, and others when they are germinating, as with bean sprouts, mustard and cress. Seed germination is also an important part of the brewing industry. Ripe barley seeds, for example, are harvested and then germinated. The starch foods stored in the grains are converted into sugars – in this case, malt – by the chemical processes that accompany germination. When all the food has been converted into malt, the seedlings are killed and dried. The malt is then fermented with yeast to make alcohol and carbon dioxide. creating the bubbles in the beer.

Plant Growth and Development

A plant that is larger and heavier than it was, for example, the day before has not necessarily grown. It may only have absorbed a great deal of water. This occurrence is similar to when a person has eaten an enormous meal. Immediately afterwards, he may weigh more, but he has not actually grown physically in size.

Plant growth takes place when the amount of protoplasm in the plant increases. This process involves the division, enlargement and development of the cells. The process is seen most clearly when the plant produces a new leaf or bud.

As soon as a seed germinates, the cells of the root and the shoot start to grow. Each cell at the tip of young roots and shoots divides rapidly to produce more cells. The genetic material in the nucleus of each cell becomes shorter and thickens to form sausage-shaped chromosomes, which arrange themselves across the center of the nucleus. Each chromosome then splits into two identical, parallel parts. These pairs of chromosomes soon separate again and travel to either end of the nucleus so that it now contains two identical sets of chromosomes. A wall then forms across the middle of the cell, producing two identical and distinct cells, each with its own chromosome-carrying nucleus and cell wall. The process may then be repeated all over again.

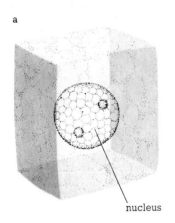

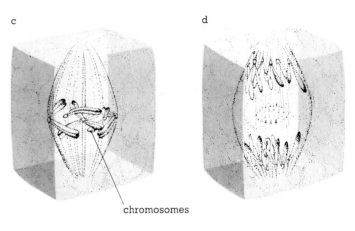

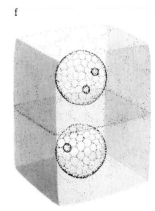

This diagram shows the reproduction of a plant cell by mitosis, starting from the resting phase (a). The chromosomes in the nucleus shorten and thicken (b), and become individually recognizable. The chromosomes arrange themselves in pairs across the center of the cell, surrounded by a long fibrous structure called the spindle (c). These pairs migrate (d) to opposite ends of the cell. At the same time, a wall starts to form across the cell and divides it into two new cells (e). The nuclei then enter the resting phase again (f). The process may then be repeated.

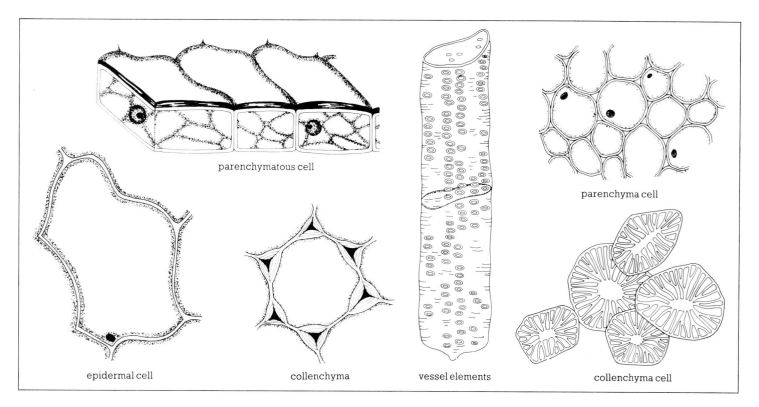

Different cells of a plant are shown above. Parenchymatous cells generally occur in all organs of a plant, while epidermal cells occur in the outermost tissue of leaves and young roots and skins. Collenchyma cells and sclerenchyma cells help support the plant. Xylem vessels carry water through the plant.

The newly formed cells grow to about the size of the original cell. In a quickly growing root or shoot the cells divide rapidly. Eventually, they stop dividing and begin to make tissues, organs and organ systems. Chemicals added to the cells cause a change in the cells' shape, a thickening of the cell walls and the formation of chloroplasts and other new cell parts. Although we know how a plant develops from a seed to a mature specimen, how it follows a correct and precise pattern of growth throughout its life continues to remain a mystery.

Plants grow throughout their lives. The genetic chemical DNA in the chromosomes controls the growth of a plant by manufacturing in a particular order a series of growth substances, or hormones, that start, stop, speed up and slow down growth. The plant also produces other chemicals that affect its growth in response to heat, light and other conditions.

In some plants that live only for a short time the growth process occurs only once. Annuals, such as marigolds and asters, live for just one growing seaon, while biennials such as foxgloves and sweet williams, start growing one year and complete their life cycle the next. Both annuals and biennials die when they have shed their seeds. Perennials, however, which include trees and shrubs, and flowers such as peonies and carnations, grow for more than two seasons. For them, growth is a continual process of leaf unfolding, flowering, seed-ripening and -shedding and leaf-fall or decay.

Plants in the cooler parts of the world usually have a resting phase between periods of active growth. Annuals rest as seeds, and perennials during their leafless period. Most plants grow less than half an inch a day, but asparagus cultivated in controlled conditions has grown as much as fifteen inches in one day.

The lifetimes of perennial plants vary greatly. Most trees live for hundreds of years, flowering annually, and plants that reproduce themselves by runners, such as strawberries, can live indefinitely. A few trees grow for fifty years or more without flowering, but when they finally flower, the entire plant dies shortly afterwards.

Plant Behavior

Plants, like people and the other animals, react throughout their lives to the environment. The conditions that affect any plant include temperature, water supply, light, air and contact with other plants of the same species, as well as with other species and animals. Within a certain range of conditions a plant will live and grow from the time of germination through body growth, to flowering, pollination and the scattering of seeds. If, however, conditions become too extreme, the plant will stop its ordinary pattern of behavior – for example, it may not reproduce – and it may, in some circumstances, even die.

Plants usually have a range of conditions under which they grow normally and a wider range under which they will survive but not reproduce. At different stages in their life cycles plants require different conditions. The germination of seeds, for example, often requires little or no light but a fairly high temperature. In contrast, a growing seedling needs plenty of light but a lower temperature. The process of flower opening requires a higher temperature and plenty of light, while the scattering of seeds usually needs even higher temperatures. Each plant can grow successfully and produce offspring only under the right conditions, in the right order and for the right period of time.

In extreme conditions a plant can take action to prevent damage and death. For example, many plants that live in cold areas develop a special type of sugar in their cells that holds water very tightly and does not release it. When the air temperature falls to freezing, a

This pitcher plant, Sarracenia purpurea, *common in North American bogs, traps insects in its water-holding pitcher, formed from a leaf.*

thin film of pure water on the inner tissues of leaves freezes and is instantly replaced by water taken from inside the cell. This water also freezes and is replaced in the same way. Without the water-retaining sugar, the cell would eventually lose all its water. The resulting increase in the concentration of salt would cause, in turn, the leaf's proteins to solidify. Eventually the cell would die. But with the sugar, the leaf's proteins do not solidify and it survives.

When plants are wounded, they create a hormone-like ethylene compound called "traumatin." It

These two photos at right reveal how watercress seedlings grow toward the light. Those in the dish on the left have been given light in all directions. Those on the right have been placed on a windowsill in a dark room and have leaned over toward the light.

The two pictures of a chamomile plant at left show how the plant responds to drought. In normal conditions the rays of the plant's flower are open as at far left. In dry weather the flower curls up.

encourages the cells surrounding the wound to grow quickly to close it and then to produce cells like those that were damaged. This process is similar to the way our bodies behave when wounded. Ethylene is used in gardening to stimulate cuttings to produce roots so that they can grow into new plants.

Various plants move in different ways, and they produce hormones that are responsible for their movement. Bacteria, single-celled water algae and certain fungi can usually move their whole bodies; other plants can move only parts of their bodies. Of course, fruits, seeds and pieces of plants are moved from place to place by wind and water currents. The movement of a part of a plant in response to an outside cause is known as a tropism. Geotropism is the response to the pull of gravity. Roots grow downward, and are said to be positively geotropic. Stems grow upward because they are attracted to light, and are positively phototropic. Tropisms can also be negative. The plant may respond to stimuli such as heat and certain chemicals by physically moving away from the harmful source.

An unusual behavior is shown by insectivorous plants. In some, as soon as an insect settles on a leaf, tentacle-like hairs close and prevent the insect's escape. Glands on the hairs secrete juices that kill and then digest the insect and absorb the nutrients released. Other insectivorous plants have vase-like leaves to imprison and digest insects. The Venus's flytrap has a hinged leaf that snaps together around the insect.

*The leaflets of the sensitive plant (*Mimosa pudica*) will fold up when they are touched. They will open out again in about fifteen minutes.*

Plant Diversity

More than a million and a half different living species of animals have been discovered, nearly three-quarters of them insects. In contrast, only about 400,000 living species of plants have been found, including all of the bacteria and viruses. Nevertheless, the plant kingdom is as diverse in shape, size and behavior as the animal kingdom. The smallest plants, such as bacteria, are smaller than the smallest animals; the largest plants, such as the redwoods and the giant sequoia, rival the whales in size.

Plants grow in diverse habitats and on every continent. Some fungi and bacteria spend their entire lives in the intestines of humans beings, where they help to digest and absorb food. Of course, most plants live freely. Flowering plants grow even on the tops of mountains, and climbing plants thrive in the depths of tropical rain forests.

To survive in a wide range of habitats and environments, plants have developed many different mechanisms. Deep-rooted plants evolved in arid regions. Plants with long, shiny leaves, which allow water to run off easily, evolved in the tropics. Many plants are able to present the maximum surface of their leaves to the sun, or produce fruits that are eaten by birds, which then disperse the seeds in their droppings.

Some plants complete their life-cycles – from seed

Today there are more different flowering plants than any other single group of vascular plants. In the past, as shown on this chart, there were many different groups of plants which were important at different times in the earth's history. The earliest vascular plants probably evolved from the green algae with the briophytes as an intermediate step. Early groups, such as seed ferns, are now extinct. Others, which are the descendants of the early groups, still survive along with the flowering plants.

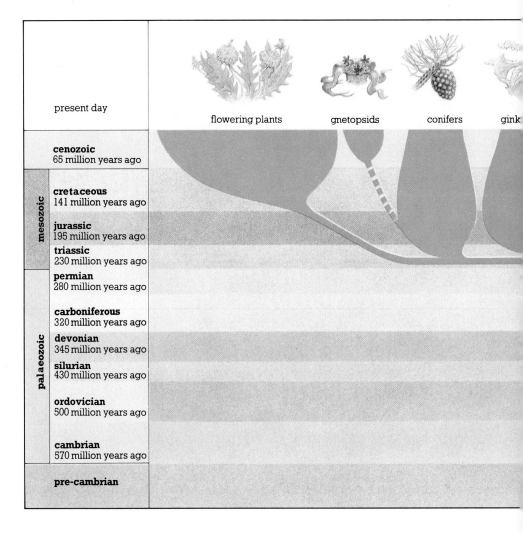

through growth, flowering, fruiting, to seed again – in less than three months. Others, such as some trees, palms and large succulents, take many years to reach maturity, flowering once, and then dying as the seeds are shed.

With only scanty fossil records to guide them, scientists believe that the earliest forms of life on earth were very simple single-celled plants that may have been like some kinds of bacteria known on earth today. Over millions of years these plants developed more complex cell parts and later evolved into multicellular plants and animals. Multicellular organisms, such as the larger forms of algae, came to dominate the early oceans and probably the land as well. Some algae, unable to manufacture their own food by photosynthesis, obtained it instead from dead and decaying plants and animals, or parasitically from still living organisms. These became the fungi. Mosses and liverworts evolved from the multicellular algae and were probably among the earliest plants to live only on land. While water was still required for reproduction, these plants could survive in places that were merely moist and damp.

The oldest land plants to be preserved as fossils date back about 400 million years and had already developed water-carrying tissues, the xylem and phloem, which allowed them to stand upright. The ferns and related plants, such as horsetails and club mosses, all of which reproduce by spores, are the closest modern day relatives to these early land plants. By far the most abundant plants today, both in number of species – well over 200,000 – and in the total number of individuals covering the land surface, are the flowering plants, or angiosperms which proliferate in almost all parts of the world.

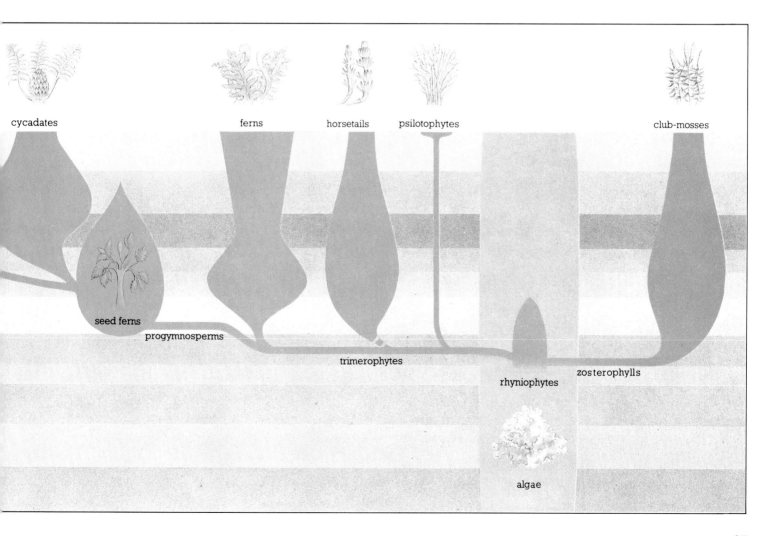

Classification and Names

Plants are classified according to features they have in common with other plants. In the eighteenth century plants were classified according to details of overall appearance, such as leaf shape and number of floral parts. Today, the plant's internal structures, chemical compositions and methods of reproduction are also taken into consideration.

The basic unit of plant classification is the species. A species is a group of individual plants that have in common definite, permanent characteristics and that usually breed with each other to produce similar offspring. Separate species that have features in common are grouped into a genus, and genera (the plural of genus) are grouped into families, families into orders, orders into classes, and, finally, classes into divisions. There are also groupings between these categories, such as subclass, subfamily, and subgenus. The formation and use of all classificatory names are controlled by the International Code of Botanical Nomenclature.

Botanists are entitled to invent and use their own classificatory schemes, and new ones are continually being published. The majority of these are only refinements and minor regroupings of older schemes. They all keep the major categories of genera, families, orders, classes, and divisions. These classifications are usually based on theories about the evolution of plants that botanists accept but cannot prove.

All such schemes, whatever their purpose, refer to species by a double name, or binomial, which is internationally accepted. The binomial is always in Latin and is given only after a detailed investigation into the plant's relationships to other apparently similar plants and its place in the scheme. The common names of plants are rarely used by botanists. Some common names are understood only in a single country or region, and others can refer to different plants in different places.

To be strictly scientific, the name of a plant should be composed of three parts. The first is the generic name, which refers to the genus; the second is the specific epithet, which identifies the different species within the genus. These two names form the binomial. The third part is the name of the botanical author who first correctly described the species. For example, *Achillea millefolium* L., popularly known as the common yarrow or milfoil, is a native European weed; forms of it are cultivated as a garden plant in North America. *Achillea* is the generic name and is given to all plants

Carl von Linné, or Linnaeus, as botanists all over the world know him, devised the system of using two Latinized words to name a plant or animal.

The engraving at left shows the nineteenth century British botanist Dr Joseph Dalton Hooker collecting plants in the Himalayas. Besides being a botanist, Dr Hooker was a noted artist whose drawings of plants, particularly of ferns and mosses, were popular collectors' items at that time.

considered similar to the common yarrow. Generic names can be based on any real, mythical or imaginary name in any language, but must be Latinized. *Achillea* is named for Achilles, the legendary Greek hero who used the plant medicinally.

Millefolium is the epithet that distinguishes the common yarrow from other species of *Achillea*. The specific epithet is usually based on a characteristic of the plant, such as its color (purple, *purpurea*), size (large, *gigantea*), habitat (water, *aquatica*), or geographical location (America, *americana*). It can also be derived from the name of a person connected with its discovery or study (Mackay, *mackayi*, or Rothschild, *rothschildiana*). In this case, the reference is to the plant's leaves: *mille*, thousand, and *folium* leaf.

The third part of the plant's name – the botanical author – is abbreviated. L. is the accepted abbreviation of the eighteenth century Swedish botanist Carl von Linné – or Carolus Linnaeus – who devised the present day international system of naming plants and animals. In general usage the names of the author are omitted, but in scientific works they are always used.

The rules governing the formation and use of names, and the transferral of a name if a plant is reclassified are periodically revised at special sessions of the International Botanical Congress held every six years.

Bacteria and Cyanobacteria

Bacteria are present everywhere – in the soil, in water, as spores in the air, and in the bodies of dead and living plants and animals. Some bacteria are germs that cause disease, and for this reason much of the research into the habits of bacteria is aimed at finding ways of curing and controlling human disease. Many bacteria-destroying drugs, known as antibiotics, have been found. The most effective of these is penicillin, which is produced by the fungus *Penicillium chrysogenum*. Not all bacteria that grow in living animals, however, cause disease. Species living in the intestines of animals are necessary to change the foods the animals eat into forms that their bodies can use.

Most bacteria are microscopic, single-celled plants that multiply by dividing into two identical cells. These new cells grow quickly until they, in turn, divide in two. This reproduction is so fast that a single bacterium can produce over thirty-two million similar bacteria in twenty-four hours. This growth of a bacterial colony, however, depends on the correct temperature. Most species will grow at temperatures between 50°F and 70°F, and some species that live in the soil in high altitudes will grow at temperatures as low as 35°F. Hot springs, and even hot water flowing from factories and electrical generating stations provide homes for a few specialized species that grow at 170°F. When conditions are not suitable for growth, for example, when there is little food or water, bacteria produce thick-walled spores that can remain dormant for several years until the environment is suitable.

Most bacteria are spherical and can be grouped in pairs, fours, chains, spirals or irregularly. Others are shaped like straight or curved rods and can be joined end to end to form long threads. Most bacteria cannot move, but a few types have one or more whip-like projections, called flagella, that they use to push themselves through the liquid in which they live.

Most bacteria take their food from other living or dead plants and animals. A few make it by their own

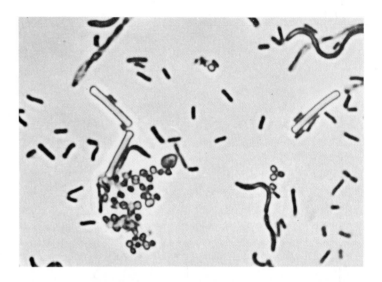

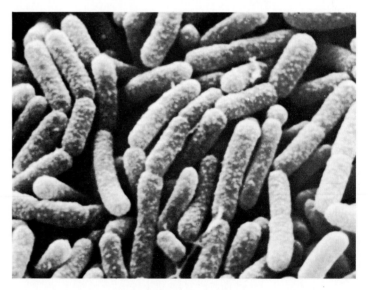

New photographic techniques enable us to see the shapes of individual bacteria in a colony. Sausage-shaped species, above, are among the most widespread of disease-producing bacteria. Colonies are often chain-like, top, but the links can easily become detached to start another colony.

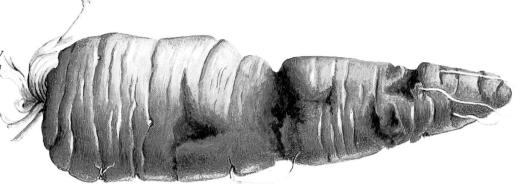

Here, a carrot has been attacked by bacterial disease that has tunneled into the vegetable's flesh. Eventually it would devour the entire carrot and turn the carrot's complex proteins and carbohydrates into simpler chemicals.

form of photosynthesis, which produces no oxygen. A few types, unique among living organisms, exist on inorganic substances such as nitrogen, iron, sulfur and ammonia. When the earth first cooled, the atmosphere contained a great deal of ammonia, and the bacteria that used this chemical for food were possibly among the first living things. Nitrogen bacteria are important because they cause nitrogen and oxygen in the air to unite chemically to form nitrates, which are essential to the life of plants. Some live in the soil. Others enter the roots of plants, causing the roots to swell and form rounded lumps, or nodules.

The bacteria that live in the soil are useful to plants and people. They help to decay dead material by turning it into simpler substances that are used again by living plants. Some of these bacteria break down solid sewage and others are used to purify water.

Similar to bacteria are the cyanobacteria, or Cyanophyta. Often called the blue-green algae, the cyanobacteria resemble bacteria in that they lack complex cell parts, a well-defined nucleus, and have no sexual reproduction. Multiplication takes place by cell division. The resulting offspring usually stay together to form a colony held together in a jelly-like envelope.

Most species of cyanobacteria live in fresh water, but some grow on damp rocks as colonizers in soil and in the outflow from hot springs. These thermal algae, as they are called, can multiply in temperatures as high as 185°F. They may have been among the first plants in the world.

Spirulina maxima, left, is a cyanobacterium similar to the earliest forms of life on earth. Today they are found in seemingly hostile places such as chemical-rich hot springs and sewage treatment plants.

Viruses and Phages

Viruses are parasites that live in the cells of plants and animals and cause many serious diseases. Crop plants, domesticated animals and humans all suffer from virus diseases. Smallpox, influenza and poliomyelitis are all caused by viruses. Viruses are so minute – as small as six ten millionths of an inch long – that they can be seen only with an electron microscope and can pass through the finest filters.

Many scientists have argued that viruses are not really plants or animals, and are not even living, because they can be crystallized and redissolved in the same way as ordinary chemicals. However, viruses can reproduce themselves and pass on their inherited characteristics, as only living things can do. Viruses are often considered plants because they are similar to small bacteria.

The individual virus is very simple in structure. It is typically composed of a protein coat around a core of DNA or RNA, the genetic material that carries information from one generation to the next. When a virus invades a cell of the living host plant or animal, it can make the cell produce more viruses instead of more of its own material. The resulting virus offspring then leave the host cell, which may die because of the takeover of its functions. The rate of virus reproduction is very fast, and a whole plant or animal can be killed by a virus in less than a day. Some viruses may lie dormant for many years before they start to reproduce. It is believed that many forms of cancer may be caused by viruses that enter a host cell and become part of the cell's own genetic material.

The virus diseases of some plants can be passed on from one generation of the host plant to another. For example, the color markings of the flowers of Rembrandt tulips are caused by an inherited virus. Virus diseases are generally transmitted either by direct contact or by means of a transmitting agent. These agents can be insects that feed off virus-containing plant and animal juices, and thus carry the viruses from one host to another. Even a gardener's pruning knife or scissors can carry infected sap from one cut stem to another. Virus diseases in plants, often called mosaics

In the diagram below a phage enters a bacterium and destroys it. The genetic material of the phage commands the bacterium to replicate the phage. The offspring of the phage then burst out of the bacterium, thus killing it.

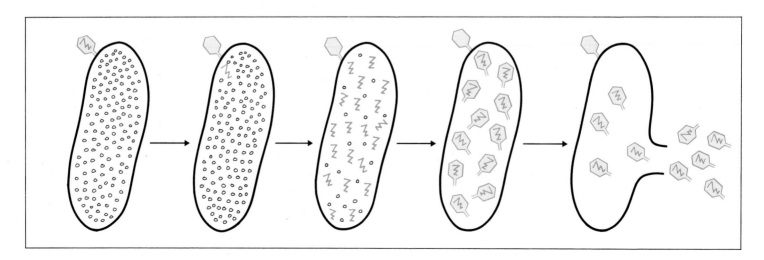

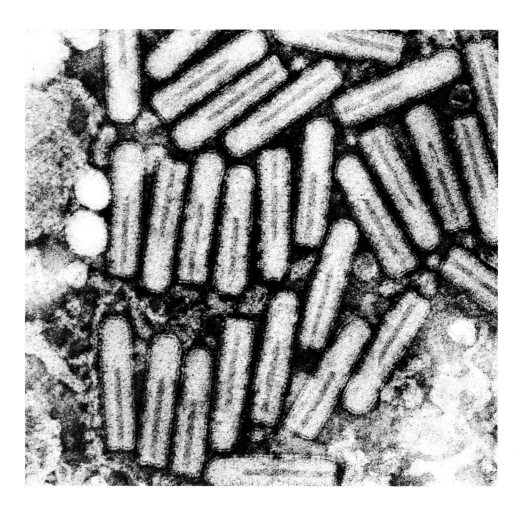

These virus particles, magnified about 100,000 times, were taken from the sap of a diseased Brussels sprout plant.

When greatly magnified, some virus particles look like the simple rod shapes, below left. Enlarged still further, below right, the protein sub-units that make up the protein coat of the virus become clearly visible.

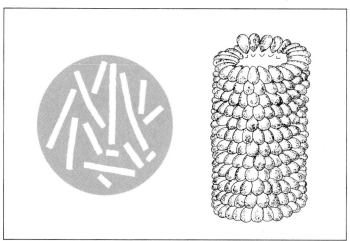

because of the appearance of the diseased part of a plant, are very difficult to eradicate. The gardener's best remedy is to burn all of the affected specimens and grow all new plants in the best conditions possible. In this way, the plant's natural vigor and defenses will usually be able to control the virus.

Some specialized viruses, called bacteriophages, or phages for short, infect bacteria. The phage consists of a protein tail and a head containing the DNA. The phage attaches itself to the wall of a bacterial cell and injects its contents of genetic material which causes the bacteria to produce more phages rather than more bacteria. Eventually the bacterial cell bursts and releases the phages, which attack other bacteria nearby. Bacterial diseases can be controlled by introducing a phage produced in a laboratory into the diseased organism when the plant is unable to provide its own defense system.

Algae I

Fossilized algae dating back more than three billion years have been found in rocks from Australia and are among the oldest known fossils. Some 20,000 species of algae exist today. Most of them are single-celled or form threads of single cells. Single-celled algae live in fresh water, such as those that form the familiar scum on ponds, and a few species live on land. But the majority of single-celled algae live in the upper layers of the sea, where they form an important part of plankton, the drifting mass of minute plant and animal life that is so important as a food for many other species of sea life, particularly fish. These algae, called phytoplankton, are the beginning of all marine food chains.

The algae are classified into six divisions. In the division Chlorophyta, commonly called the green algae, are about 7500 species, encompassing a wide variety of forms and habitats. One species grows on tree trunks, wooden fences and brick and stone of all types; others thrive in snowbanks or on the shells of turtles. However, most species of Chlorophyta live in fresh water, and a few others live in the sea.

All of the green algae make carbohydrates from carbon dioxide and water by photosynthesis. They are an important part of the food chain for fresh water animals. Because they reproduce rapidly, and both sexually and asexually, they exist in enormous quantities, with tens of thousands of individuals of one species closely packed together in a colony.

In the division Chrysophyta are more than 6000 species, most of them microscopic. They are the yellow-green and golden-brown algae, and most importantly, the diatoms. The yellow-green and golden-brown algae can be solitary or united into colonies. They live in the sea or fresh water, and also in damp

A thread-like alga, Spirogyra, *top left, is composed of single cells joined end to end with spiral chloroplasts inside.* Volvox, *top right, is a colonial algae made of a single envelope of interconnected cells embedded in a jelly-like mass.* Licmophora flabellata *cells, bottom left, remain closely attached for most of their lives. The hard and resistant cell walls of sea-living diatoms, bottom right, are made of silica that persists for millions of years after the living cell has died. The markings are different on each species of diatom and it is easy to identify which ones lived in different places in the past.*

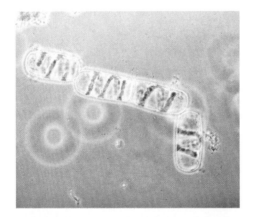
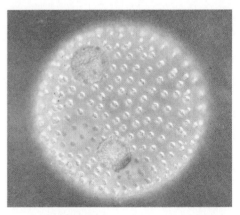

Millions of microscopically small single-celled green algae can be found on damp tree trunks, even in heavily polluted inner-city parks. The alga shown here on an old birch tree is a species of Pleurococcus. *It has the ability to survive long periods of drought by secreting a protective wall over its cells to keep it moist.*

spots, where they often intermingle with mosses, liverworts and other kinds of algae.

The diatoms are microscopic organisms that make up a large proportion of the phytoplankton in the oceans. The beauty and variety of these minute, water-living plants is amazing. The most characteristic feature of the diatoms is that the outer layer of each cell is impregnated with a hard mineral called silica, giving each species its distinct appearance. This stiffened cell wall is like a box of two almost equal parts, one overlapping the other. This unique structure accounts for the name diatom, which means two atoms. The silicanized walls are arranged either in a circular, radiating pattern or in two symmetrical parts. The cell walls remain when the living algal portion dies and decays and may be preserved in deposits that form at the bottom of the pond or sea in which they lived.

Some 5500 species of diatoms are known so far, some of them only as fossils. Vast deposits of diatom fossils are found in places that were once warm seas. Many of these areas of diatomaceous earth, as they are called, lie near oil deposits. In the Santa Maria oil fields of California, deposits of fossilized diatoms reach a depth of more than 3000 feet.

As a result of the exploration for oil, much has been discovered about diatomaceous earth and how it can be used. Scientists have found, for example, that it is highly absorbent. The first dynamite was made from diatomaceous earth soaked in the liquid explosive nitroglycerine. Diatomaceous earth is also used as an insulating material because it can withstand temperatures of over 1500°F.

Algae II

Brown seaweeds, such as Fucus vesiculosus, *right, can survive out of sea water during low tides. At high tide the seaweeds are submerged and absorb water that enables them to continue their life processes. The brown seaweeds are usually attached by their holdfast to rocks. The green alga in the picture is the sea-lettuce* Ulva lactuca, *a fragile-looking plant that is nonetheless able to survive the pounding of the waves.*

Most of the species in the division Pyrrophyta are dinoflagellates, another important component of marine phytoplankton. The dinoflagellates are single-celled, have flagella that help propel them through the water and are particularly common in the warmer parts of the ocean. Some are luminescent and in sufficient numbers can cause the sea to glow at night. Other species, called armored dinoflagellates, have sturdy cellulose cell walls arranged in plates, which resemble medieval armor. Occasionally, dinoflagellates grow so numerous that they color the sea red or bright yellow. Several species give rise to the so-called "red tide" that kills fish and other marine organisms.

Like dinoflagellates, most of the other species of Pyrrophyta are single-celled and have flagella. One species, *Dinamoebidium varians,* resembles the amoeba, which is an animal. Other members have their cells joined end to end to form branching threads.

Members of the division Euglenophtya are mostly unicellular and are commonly found in fresh water, such as barnyard pools. Euglenophytes are unusual in that zoologists have claimed that most of them are animals. They base their claim on three reasons. First, members of Euglenophyta move through the water with their flagella. However, since members of Chlorophyta and Pyrrophyta can also move in the same way, this reason by itself is not good enough. Second, Euglenophytes have a light-sensitive red eye spot, which enables them to seek out light and photosynthesize at the fastest possible rate. Third, many of them can eat solid foods. Most scientists consider Euglenophytes to be plants, however, because they produce food by photosynthesis.

Most of the single-celled algae belong to the first four divisions and are important ecologically because they are often the first colonizers of open water and damp surfaces. Some of the species grow in and help to purify sewage-polluted water.

The divisions Phaeophyta and Rhodophyta are overwhelmingly made up of seaweeds. Unlike the other algae, seaweeds are relatively large plants. One species, the giant kelp, grows deep in the Atlantic Ocean and produces a repeatedly branched stem as long as 200 feet, making it one of the largest plants.

Most seaweeds reproduce sexually by a fairly simple process, which varies slightly from one species

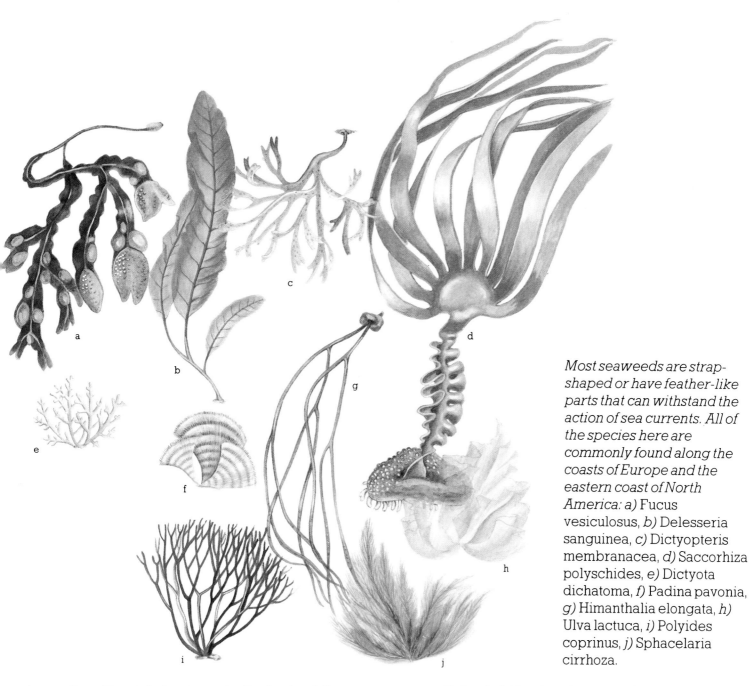

Most seaweeds are strap-shaped or have feather-like parts that can withstand the action of sea currents. All of the species here are commonly found along the coasts of Europe and the eastern coast of North America: a) Fucus vesiculosus, *b)* Delesseria sanguinea, *c)* Dictyopteris membranacea, *d)* Saccorhiza polyschides, *e)* Dictyota dichatoma, *f)* Padina pavonia, *g)* Himanthalia elongata, *h)* Ulva lactuca, *i)* Polyides coprinus, *j)* Sphacelaria cirrhoza.

to another. Special structures in the body of the seaweed produce male and female cells. These can be superficially identical or differentiated into sperm and egg cells. As in other types of plants the female cell is fertilized by the male and the resulting embryo develops into another plant.

Seaweeds are feathery or ribbon-like plants having no vein and root systems, leaves or woody parts. Instead of a root, many have a special organ called a holdfast. With its sucker-like outgrowths, it enables the seaweed to remain anchored to rocks in the roughest of seas. If the holdfast breaks away from the rock the seaweed drifts off and dies. Species that lack a holdfast live floating through the water.

Seaweeds are classified according to their technical features, which are correlated to the plant's color, so that members of Phaeophyta are known as the brown seaweeds and members of Rhodophyta as the red seaweeds. Although they can grow together, they are usually found in different zones depending on sea depth and light penetration which will determine which type is found growing where.

Fungi I

A piece of stale bread covered with small, bluish-green growths is exhibiting one of the many species of fungi. Fungi (the plural of fungus) are simple plants that, like the algae, lack true roots, stems and leaves. The body of a fungus consists of fine branching threads, called hyphae, that are constantly feeding and growing. Like a cobweb, they form a network over and usually in the food on which they live. Some hyphae are divided into cells, but often have no dividing cell walls.

Fungi have no chlorophyll in their cells and therefore do not make food by photosynthesis. One type of fungi, called saprophytic fungi, takes its food from dead and decaying plants and animals. A second type, the parasitic fungi, get their food from the tissues of living plants and animals. Both types help maintain the balance of nature. Saprophytic fungi help to break down dead bodies, leaves, old wood and the waste matter of plants and animals into simpler chemicals that can be reused by plants. Parasitic fungi cause diseases and thus control the numbers of most living things.

Saprophytic fungi, however, can also be harmful and destructive. Several species grow on leather and natural fibers, such as cotton, and secrete enzymes that

When yeast cells are photographed with polarized light through a yellow filter, above, the individual cells and their contents can be clearly distinguished enabling scientists to study their behavior.

Spores of the fungus Penicillium, *right, are widely distributed in the air. They germinate only if there is sufficient warmth, moisture and food present, but require no light. A species of* Penicillium *is grown to produce the antibiotic drug penicillin. The spores of* Penicillium *are released from the tips of the hyphae and can remain alive for many years in soil and dust.*

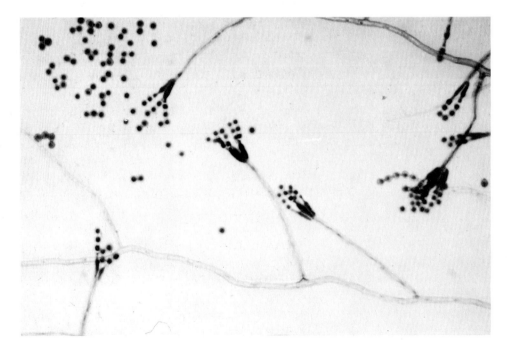

destroy the materials. Dry rot fungus attacks wood and can be a nuisance in old buildings. Some fungi attack prepared foods, a few live on gasoline and reduce its performance in cars, still others attack paint and a number of fungi can quickly destroy stonework and bricks both inside and outside the home.

There are useful fungi, too. Some molds are used as flavorings in blue cheeses. The mold *Penicillium chrysogenum* is processed to make the antibiotic drug penicillin.

Fungi reproduce sexually by the fertilization of sex cells or asexually by the production of spores. Fungus spores are very small – a thimbleful of soil can hold 100,000 of them.

Fungi – of the division Myxomycota – are divided into several classes. The slime molds, or Myxomycetes, live on decaying plants and animals. Their slime is a jelly-like mass of living protoplasm that moves slowly as it feeds. Eventually it takes a stalk-like shape and produces spores.

The Oomycetes cause many plant diseases, such as potato blight, and the damping-off of seedlings. The parasitic genus *Saprolegnia* of the class Oomycetes attacks fish, especially fish in home aquaria. Species of Zygomycetes produce the black and white molds found on jam and bread.

If the fungi-caused plant diseases of the world's crops were eradicated more than enough food would soon be available to feed the entire human population. On the other hand, without these diseases many wild plants would grow unchecked and overpower other useful species, which could even become extinct. Nevertheless, each year millions of dollars are spent on plant pathology research into the prevention and cure of fungus diseases.

Another class is the Ascomycetes, which includes *Penicillium*, another mold from which LSD is made, and the mold and the mildews found on many damp materials. This class also includes the yeasts, which are single-celled and have the ability to convert sugar into alcohol and carbon dioxide by fermentation and are used in baking and brewing.

Some of the microscopic fungi are so important in science and industry that special culture collections, or gardens, of them are maintained by university laboratories and drug companies.

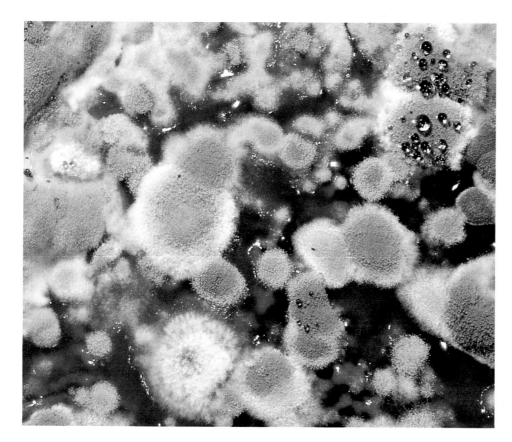

Mold-producing fungi will grow on any material if organic food and moisture are available. The fungi at left are growing on elderberry wine. When foods and drinks are stored for future use, fungus spores already present must be killed by heat or chemicals. Canning, freezing and bottling all prevent fungus growth.

Fungi II

An old wives' tale, which many people still believe, says that mushrooms are edible and toadstools are poisonous. To botanists, however, toadstools are only different names for the same plants – members of the class Basidiomycetes – the large fungi with conspicuous, spore-producing bodies. The most familiar mushroom is probably the cultivated field mushroom sold in foodstores. Like most mushrooms, it grows in soil or a special compost and pushes its body up when the weather is damp and cool. The cap and stalk of the mushroom are a mass of closely packed hyphae. The hyphae on the underside of the cap form delicate, thin gills; spores are produced in large numbers on the surface of the gills.

Altogether there are thousands of species of mushrooms, both harmless and poisonous. They are distinguished by differences in such features as the shape of the gills, the color of the spores and the color and markings on the skin on the top of the cap. In some poisonous species this skin is brightly colored, such as in the scarlet-capped fly agaric. Color, however, must never be taken as the only indication as to whether or not a fungus is safe to eat. The most poisonous North American mushroom, the destroying angel, is pure white and when young, very similar to the common field mushroom.

Mushroom poisons are complex chemicals that affect the chemistry of the animal that eats them and often lead to death. The nature and action of many mushroom poisons is not yet understood by scientists. Apparently, some animals can safely eat certain mushrooms that would be deadly to others.

A stroll in a damp woodland or grassy field will reveal many kinds of large fungi. Not only will ordinary stalk and cap fungi be seen in great quantities, but many others that look quite different. One species, for example, looks like pieces of discarded orange peel. Another resembles skinned rabbit's ears. Some fungi could be mistaken for cauliflower, others look like miniature versions of reindeer horns, and still others

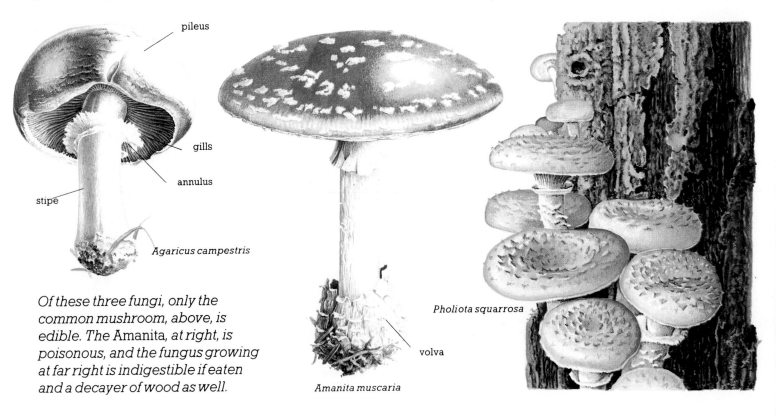

Of these three fungi, only the common mushroom, above, is edible. The Amanita, *at right, is poisonous, and the fungus growing at far right is indigestible if eaten and a decayer of wood as well.*

Agaricus campestris

Amanita muscaria

Pholiota squarrosa

Peziza aurantia, a specialized member of the Ascomycetes, left, the orange peel cup fungus, which grows on rotting timber among the mosses of the forest floor. The spores are produced on the downy inner surface. It is not a true mushroom.

The parasol mushroom Macrolepiota procera, a member of the Basidiomycetes, below, grows in damp fields, along highways and on the edges of woodlands. It is large – up to eighteen inches tall – and perhaps the most tasty species in the world.

like jugglers' clubs. Puffballs resemble large brown or gray pebbles. When they are stepped on, they emit a puff of khaki-colored spores. One species of puffball can grow to more than two feet wide, and more than once a large colony of them has been reported to the police as objects from outer space.

Probably the most unusual fungi are the bird's nests, which appear as thimble-sized brown cups on rotting wood. When the cup opens, the spore containers inside look exactly like white bird's eggs. Earthstars have ripe, round bodies that fold back in sections and press against the ground in the shape of a star. In earlier times, legends grew around these fungi which were believed to have sprung from the dust of shooting stars.

The stinkhorns are easily detected by their overpoweringly offensive smell. Hyphae growing underground produce a so-called witch's egg that bursts and releases a long, white horn with a slimy olive-green tip carrying the spores. The rotting-flesh smell soon attracts flies that feed on the slime and carry away the spores.

Lichens

Certain plants live together in a special relationship whereby each species gives something – such as food, water or protection – to the other that it cannot provide for itself. This relationship is called symbiosis. A good example of symbiosis is a fungus living in the roots of a tree. The fungus passes water and minerals from the soil to the roots and in return takes food from the tree.

Lichens are composite plants in which a certain type of fungus and one or another species of green algae or a cyanobacterium live together, with both benefiting from the association. The fungus obtains food from the photosynthetic green plants, and they obtain mineral nutrients, water and protection from overexposure from the fungus. Lichens can exist where neither component could live independently.

The body of a lichen can be like a thread, a jelly, a crust, a leaf or a tiny shrub. It can be upright or dangling, branched or unbranched. Some species are small, such as the black pin-sized encrusting lichen found on tide-level rocks. Certain lichens live in colonies that look like twisted masses of branched threads and can be more than 20 feet in diameter. In some lichens the body is simply the hyphae of the fungus in which cells of algae are entangled. More complex lichens have alternating layers of fungi and algae put together like a sandwich. Some hyphae in the lowest layer act as roots, anchoring the lichen to the soil, rock, leaf or tree-bark on which it lives.

Lichens of the genus Cladonia *are found throughout the world growing on damp bark, peat, soil and many other natural substances. They are most abundant in Arctic and temperate regions where a single species can cover a large area. The species at right is growing on the trunk of a tree in an Alaskan forest.*

Cladonia cristatella, *left, is a lichen that grows profusely in the United States. At the tips of these specimens are the spore cups. They are formed by the fungus partner spores that germinate and capture algae to form a new lichen plant.*

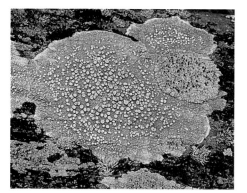

Lecanora calcarea, *far left, is a crustose, or encrusting, lichen. Other species are often found with* Lecanora, *competing with it.* Usnea florida, *left, grows on the twigs and branches of trees throughout the world. The cup-shaped, spore-producing structures, or apothecia, are larger than in any other species of lichen.*

Lichens reproduce in several ways. Small pieces of the body can break off and grow into new plants, or masses of algae cells and fungi hyphae can group to form special bodies that are carried away from time to time. Lichens can also reproduce sexually by means of fungus-type spores that germinate and form a lichen if the right algae or cyanobacteria are present.

Lichens are found on every continent, and 16,000 species have been documented. Millions of individual plants of any single species can grow in a small area. All kinds of environments are suitable for lichens, from the leaves of the trees in tropical rain forests to the bare rocks of the polar regions. In the Antarctic, lichens are by far the most diverse and abundant plants, and in the Arctic tundra the so-called reindeer moss, the principal food of reindeer, is really a lichen. In Iceland people eat a species of lichen, and in parts of the world lichens are used to dye wool. The substances in lichens that produce color are extracted to make dyes. Some lichens can be identified by noting the color they turn when certain chemicals are dropped on them. Litmus paper, which turns red in acids and blue in alkalis, is made from lichens and is used throughout the world to test the acidity or alkalinity of substances.

Lichens can survive extremes of climate. They colonize bare rocks, concrete, bricks, asbestos shingles and even glass, breaking all these materials down into small particles to form a soil. One thing that they cannot tolerate, however, is air pollution although some species are more resistant than others. Counting the number of species within various distances of city centers can determine pollution concentration.

Mosses and Liverworts

Mosses and liverworts – the division Bryophyta – are small, nonwoody, flowerless plants with no true roots, stems, leaves or well-defined system of veins. They grow in all kinds of habitats – the tops of brick walls, the forest floor, the branches of trees in the high-altitude tropical rain forests, and in the Arctic and Antarctic tundra. Many species are tolerant of drought, appearing to be dead after a long, hot, dry season but, when moistened, becoming green again and continuing to grow. This ability to survive in harsh conditions means that they are often among the first colonizing plants on newly exposed rocks and other fresh surfaces. In most cities and towns small spaces in sidewalks and crevices in stone and brick buildings are filled with the minute silvery-gray "leaves" of these pioneers.

As colonizers, mosses and liverworts are ecologically important, because they prepare sites for larger plants, but few of them are of direct use to people. The exceptions are the peat mosses in the genus *Sphagnum*, which are extremely absorbent and are used in horticulture as an ingredient of potting composts and also for packing delicate plants for transit.

Where soil is sparse, such as in a boulder-strewn woodland in the Scottish Highlands, right, mosses and lichens are often the only plants besides trees and shrubs to grow. Mosses often form a continuous blanket on the ground.

Many liverworts have a flat, shiny green thallus from which the reproductive structures arise. Marchantia polymorpha, *left, has distinct umbrella-like structures that release spores into the air. It is found on damp soil in the wild and as a weed in gardens and near paths and walls.*

Species of Polytrichum *are among the more conspicuous mosses in the northern hemisphere.* Polytrichum juniperinum, *below, grows as a turf among heather and grasses in dry, impoverished areas, usually in highlands.*

Mosses and liverworts absorb water and mineral nutrients through their leaf-like scales. Their slender root-like hairs, called rhizoids, also take in water, but serve to anchor them to the surface on which they grow. All mosses have a stem-like structure, called a seta, which in some tropical species is up to three feet long. Superficially, most of the liverworts look like mosses, having scales and a seta. The largest liverworts, however, look something like seaweeds, having flat bodies with rounded lobes, except that liverworts are mostly a bright, shiny green, rather than brown. This type has no scales or a seta, and the lobes are held to the ground by rhizoids.

Liverworts and mosses reproduce by a method of alternation of generations similar to that of ferns. In mosses and liverworts both generations of plants – spore-producing and gamete-producing – grow together (in ferns, they are separate). The moss or liverwort plant that is seen growing is the sexual form, which produces the gametes. When the male has fertilized the female, the fertilized egg develops into the nonsexual spore-bearing plant, which grows on a short seta above the main plant. In liverworts, the spore capsule eventually appears at the top of this plant. When the spores are mature, the capsule bursts and sheds the spores. If the spores fall on moist soil, they will germinate and develop into new liverwort plants.

Some liverworts can also reproduce asexually by growing small, detachable sections, which are washed out by splashing rain and soon grow into new plants.

When moss spores have ripened, scattered and germinated, they develop into a very fine, hair-like green plant from which clumps of typical mosses grow.

Although mosses and liverworts require water for reproduction, they are mainly land plants. Descendants of the early, primitive water-requiring algae, they are in a separate evolutionary line from plants with vein systems, such as ferns and flowering plants.

Ferns and Their Relatives

Ferns and their relatives are simple plants that have no flowers and reproduce by spores rather than seeds. Although there are several other groups of flowerless plants such as mosses, fungi and algae, ferns are unlike them in that they have a system of veins and strengthening tissues similar to those found in the gymnosperms and flowering plants. Ferns of all kinds were much more abundant in the past than they are today, and many of their fossilized remains are found in various rocks. These fossils are scientifically valuable because, by doing detailed studies, scientists gain insights into how plants evolved and gave rise to the flowering plants so abundant and widespread today.

The 10,000 or so species of living ferns so far recorded have been classified in many different ways. Although many different ferns are superficially similar in appearance, their features vary, for example, in the number of times each frond is divided into progressively smaller parts, and the position of the spore-bearing organs, or sporangia. In some species the sporangia are borne on separate shoots quite unlike the fronds. In other species the sporangia are carried on the fronds themselves, usually on the underside. The sporangia can be arranged in a regular pattern, but are more often randomly scattered.

Ferns are found wherever moist conditions exist, such as forests and marshy areas. But if the air is humid enough, several species will root in crevices on bare rocks. One group of ferns lives only in water, and in tropical countries, one species, *Salvinia melesta*, has become a major pest in reservoirs and artificial lakes. Although ferns require the same conditions for growth as flowering plants, most can tolerate lower light levels and therefore make good houseplants.

In the tropics some ferns grow from a trunk-like stem with fronds at the top, and the whole plant, called a tree fern, looks like a small palm tree. In prehistoric times, tree ferns were common; today they are much rarer and can be found only in New Zealand, and South and

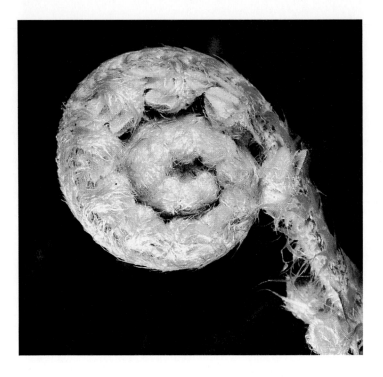

A fiddlehead of the male fern Dryopteris filix-mas *will quickly uncurl as the spring warmth stimulates growth.*

Central America as well as in botanical gardens.

Closely related to ferns are the club mosses and horsetails, which were the dominant plants millions of years ago. Few species of these primitive plants are found today, but some have managed to compete with flowering plants and have become weeds, such as the field horsetail. Club mosses occur in both temperate and tropical regions and are most spectacular in the tropical rain forests, where they festoon the lower branches of trees.

There are several other groups of fern allies. One genus *Selaginella* has about 700 different species, most of which grow in the tropics. *Selaginella kraussiana* belongs to this family. A trailing plant, it is a native of

Looking like a small fir tree, the wood horsetail, or Equisetum sylvaticum, *rarely grows more than eighteen inches high.*

The nearly uncurled fronds of the Hart's-tongue fern, Phyllitis scolopendrium, *are here seen growing in an English woodland.*

The sturdy bracken Pteridium, *above, covers the ground in pinewoods where few other plants can survive.*

Africa that now grows in the warmer areas of Europe and is commonly cultivated in greenhouses. The completely aquatic quillworts are rare but grow in all parts of the world. Other plants related to ferns grow on damp rocks or the bark of trees in the tropics and subtropics.

One hundred years ago, when a "pteridomania" craze swept Europe, odd, distorted forms of wild ferns exchanged hands for large sums of money. Today, ferns are great favorites for garden and greenhouse cultivation, and make attractive houseplants.

Palms

Palms are the only group of trees among the monocotyledons. Palms are a very large family and are important for both ecological and evolutionary reasons. They are even more important for economic purposes – the economies of many countries depend on palms as a source of export. Palms are not true trees, such as pines, spruces, willows and birches, but tree-like plants, members of the family Palmae. They are some of the few tree-like plants among the monocotyledonous flowering plants. Monocotyledons, or monocots as they are sometimes called, are plants in which the embryo within the seed possess only one seed leaf. Yet most of the 3000 species of palm resemble trees because they often have a single trunk topped by a crown of leaves. Some species of palm, however, are shrub-like, several are climbers and a few are herbaceous.

Palms differ from trees in that no true wood is produced in the trunk. Instead, the palm's rigidity is maintained by strong fibers wrapped around xylem and phloem tissues that extend upward as the trunk grows. Few palms produce branches. In most, all growth is concentrated at the top of the trunk. When the old leaves in the crown die, their bases form the characteristically scarred bark-like surface of the tree.

The leaves of the various palm trees appear different, but they are, in fact, of two types only: fan-shaped, or like huge feathers growing from a single stem. The flowers of palms are varied. Some species produce simple spikes, and others bear as many as a quarter of a million florets, or modified flowers, at one time. The effort to produce these florets can be so great that the plant dies as soon as the seeds are formed. An example of such monocarpic, or one-time flowering palms, as they are called, is the vigorous sago palm. It takes nearly twenty years to reach flowering size – only to die a few months later.

Most members of the palm family are natives of the tropics and subtropics, but a few species grow in temperate regions. The Asiatic Chusan (*Trachycarpus fortunei*), or dwarf palm, thrives in gardens and parks in coastal areas of Europe and North America. Some

The date palm, Phoenix dactylifera, *right, is widely grown in the Middle East and throughout North Africa. It is a staple food in countries such as the Sudan. Dates are exported to other countries where they are considered a delicacy.*

Elaeis guineensis, *the oil palm, above, produces large quantities of fruits from which palm oil is extracted. It is used to make soap and also as an industrial lubricant.*

Coconut palms, Cocos nucifera, *thrive near the sea, as on this plantation on St. Vincent in the West Indies, left. Most of these coconuts are used locally, but when dried they can then be exported as a valuable source of edible oil.*

palms are tolerant of drought and are often the only trees found in desert oases. Others are resistant to salty air and live on beaches close to high-tide marks. In tropic regions, palms often grow at the edge of clearings. Many species are also cultivated in plantations and yield a range of important crops.

The coconut palm (*Cocos nucifera*) provides the well-known nut kernel that can be partly dried to make edible coconut or oven dried to give copra, which, when pressed, gives an oil used for margarine and other products. The coconut's milk is drunk and its outer husk fibers are used for matting. The date palm (*Phoenix dactylifera*) is a major source of food that is exported all over the world. The oil palm (*Elaeis guineensis*) from western Africa produces an edible oil that also has many other use, such as for soap, lubricating oil and tallow for candles. Sugar and molasses are refined from the sap of sugar palms such as *Arenga saccharifera* of Java, and sago meal is made from the spongy tissue from the center of the trunk of sago palms such as *Metroxylon sagu*.

Many palms are grown as houseplants or in greenhouses, both for their ornamental leaves and also because they tolerate long periods of neglect. Palms that grow near the sea are valuable to geographers and oceanographers. Their large seeds, such as the coconut's, can be carried great distances by sea currents and then germinate in a new location. By mapping the distribution of various palm species, scientists have been able to determine past patterns of sea flow.

Habitats

Wherever on the earth's surface there is warmth, light, air and supplies of mineral nutrients and water, plants and animals can live. An area in which these conditions exist is called a habitat. The conditions affecting plant growth in a habitat are either physical, those of climate or geography, or biotic, the effects of plants on each other and the effects on them of animals and man.

Climatic factors include precipitation, temperature, humidity, wind and light. Geographical factors include topography, drainage and soils. These factors act together, but in many habitats one factor has an overriding influence, such as the low temperatures in tundra habitats or the salt in the air and the soil of coastal areas. Precipitation is the term used to describe all water that reaches the earth's surface as rain, snow, hail, frost and dew. The aspects of precipitation most important to a habitat are the average annual rainfall, its characteristic form – gentle showers or deluges – and its distribution throughout the year. The aspects of temperature that influence plant growth are the monthly average temperatures and the annual and daily extremes. Many plants require a set sequence of temperatures for set periods of time to successfully complete their life-cycles, from seed germination and seedling growth through shoot growth, flower development and pollination, to seed ripening and dispersal. Some plants require different daily periods of light in a set sequence to enable them to complete their flowering. Light intensity is important in controlling photosynthesis and developing flower colors.

Soil features that control plant growth are acidity or alkalinity (known as pH), depth, fertility and air- and water-holding capacity. Land surface features matter too by affecting still other factors, such as retention of water in the soil, rainfall run-off and the angle of the soil in relation to the sun's rays. For example, high mountain ranges can cause shadow deserts and act as barriers to plant migration.

There are thousands of different habitats in the world but they are not classified by any universally accepted system. Instead, scientists name habitats after obvious or dominating factors. So they speak of rain forests, wetlands, grasslands and woodlands. Habitats are further characterized by the plants that grow in them.

In a beech wood, left, the dense shade cast by the beeches, together with a thick layer of slowly decaying leaves, allows for little other plant growth. Where light does manage to penetrate, a fallen sapling produces a few thin leaves before dying.

At high altitudes, right, mountain lakes form on watertight rocks. Little plant growth occurs because the waters in these lakes are unusually acid, low in nutrients and cold for most of the year.

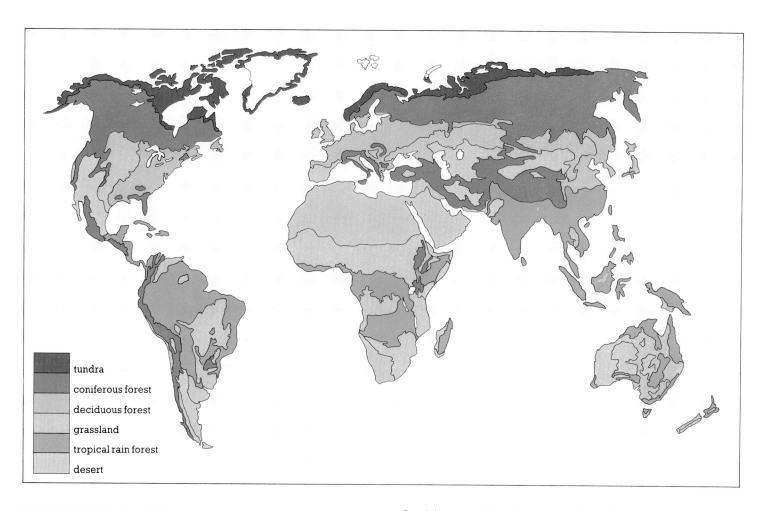

On this map of world vegetation, the main zones designated by botanists are closely linked with climate, especially temperature and rainfall.

- tundra
- coniferous forest
- deciduous forest
- grassland
- tropical rain forest
- desert

Normally, habitats change as their communities of plants change. Each colonizing plant modifies the habitat, perhaps making it more suitable for some plants and less suitable for others that grew well earlier. These changes create a succession of plant communities and, eventually, when the habitat can no longer be changed by the plants growing in it, and the climate remains the same, the habitat may be marked by what is called a climatic climax community. Natural successions end in climax forests, except in very dry areas where they become savanna. The activities of man and animals can alter a habitat, and a severe disturbance such as a volcanic eruption, can affect plant growth drastically for a while. Long-term upheavals – such as an ice age – bring permanent change to a habitat.

Wetlands

A wetland is a habitat in which a continuing supply of water is the major factor controlling the diversity and quantity of plants and animals that live there. Marshes, swamps, bogs and coastal habitats such as salt marshes are all wetlands. Of all the world's habitats, the wetlands are one of the more threatened by man's activities. Drainage, water pollution, trampling, grazing and even the destruction of entire wetlands have caused the extinction of many wetland plants, and many more will be extinct by the end of this century. Thus conservation of these areas is of the utmost importance.

All plants require some water for food manufacture and growth. Some, like ferns, mosses and liverworts, live on land but still need water for reproduction. Others, like most algae, spend their entire lives in water. Although most flowering plants live on dry land, many thrive in marshy areas, and hundreds of species remain more or less submerged for most of their life cycles. Most of these water plants, or aquatics, are specialized and have flexible stems that bend easily in water currents. The leaves of many of them are finely divided and hair-like to reduce friction with the water.

The pollen of water plants can be carried by insects but in some species the pollen is carried by water currents and caught by trailing underwater stigmas. Some water plants reproduce through special buds that break off from the main plant and fall to the bottom,

Many plants evolve to fit their natural environments. The ribbon-like leaves and elongated, spiraling female flowers of the freshwater eelgrass, Vallisneria spiralis, *above, enable it to withstand strong water currents.*

The toad rush, Juncus bufonius, *right, grows in ditches and damp pathways. Although it is not a vigorous plant, it reproduces quickly from the seeds it produces throughout the growing period. In the tussock sedge,* Carex stricta, *far right, each tussock is a single plant growing on dead plant remains.*

Marram grass, Ammophila arenaria, *above, continues to grow even when covered in sand. It uses its network of roots and remains to bind sand particles and in this way stabilizes sand dunes along coastlines.*

On this seashore large areas of shingle beach support species such as the sea-cabbage, Crambe maritima, *shown here in bloom.*

where they remain throughout the winter. These buds start to grow in the spring when the increasing amount of daylight turns the buds' stored food supply into oils which force them to rise to the surface of the water, and they begin to grow into new plants.

Most water plants are rooted in mud or attached to boulders or gravel at the bottom of the wetland in which they grow. Some water plants have only underwater leaves, others only floating leaves and still others have both kinds. Some water plants that have their roots and main stems below the water have leaves extending above the water surface.

The leaves on a water plant may all be of the same type or of different types depending on their location on the stem. In times of drought, when the water level falls, some plants survive on the drying mud and produce land-plant type leaves. Plants that ordinarily produce floating leaves or leaves that reach above the surface of the water will stop producing them if the water currents become too strong and will then exist only on their underwater leaves.

Waterlogged soil, in which the water level is at or near the soil level, can also be a habitat for plants. One type of waterlogged soil is found in peat bogs, where a hardpan under the soil allows acid water to accumulate, which prevents the total decay of dead plants. The partially decomposed remains of plants are preserved by the acid water just as cucumbers and other vegetables are pickled by vinegar. Each year a new layer is added, tightly compressing the lower layers. The result is what is known as peat, the upper layer acting as soil for living plants. Peat is poor in nutrients and therefore the plants found in bogs are either mosses, which require small amounts of minerals, or sundews, pitcher plants and other flowering plants that get their nourishment by trapping and consuming insects.

Marshes usually occur on soils rich in minerals and decaying plants. Because the water in marshy soils is not acidic, peat does not form. Marshes have many more plants than bogs or swamps. The waterlogged soil, however, means that the roots of these plants may be deprived of air. For this reason, many marsh-dwelling plants have developed large internal air bladders that take in air through the stems and pass it to the roots.

Tundra and Alpine Plants

The North and South Poles and the immediate surrounding areas have climates totally unfavorable to plant growth. Except for some bacteria and lichens, life cannot exist where there is constant daylight for half the year, nonstop night-time for the other half, continual high winds and temperatures rarely above freezing and often as low as –100°F.

Farther away from the Poles, however, the ground surface is not permanently frozen. This is the tundra, and here as many as 1700 different plants grow.

In the tundra the growing season between two cold periods can be as short as three or four weeks. Thanks to the low average temperatures, dead tundra plants decompose slowly and the accumulation of these partly decayed remains forms peat, which covers the ground like a soggy blanket. Flowering plants grow either on the peat or the coarse, gravelly debris of rocks shattered by frost.

Large tracts of tundra are found all around the Arctic Circle, northern Canada, northern Scandinavia, Siberia and along the coasts of Iceland and Greenland.

Soldanella alpina *is a typical example of alpine flowering plant, being short-stemmed with brightly colored flowers. As the snows melt in spring, it turns its flat leaves towards the sun.*

The bright red leaves of a flowering plant contrast with the white and gray patches of lichen colonies on the barren, rocky surface of this example of a tundra habitat in the far north of Canada.

Found throughout the Arctic and high mountain areas of the northern hemisphere, Dryas octopetala forms large colonies. It is usually associated with other Alpine-Arctic species such as dwarf willows.

In the southern hemisphere only some coastal areas of Antarctica have patches of tundra. The Antarctic tundra has only four species of flowering plant, but there are more than 500 in the Arctic.

Tundra vegetation is dominated by mosses, lichens and grasses. Their rate of growth is slow, and when the temperature falls after the short summer these plants sometimes remain frozen until spring, ten or eleven months later. Herbaceous flowering plants of the Arctic, such as species of forget-me-nots, willow herbs and veronicas, grow as flattened rosettes, which can withstand high winds and present the maximum leaf surface to the sun. These plants either overwinter as seeds or can survive when frozen or covered by snow. *Cochlearia*, a kind of scurvy grass, is frequently frozen when in full bloom. After thawing the following spring, the flowers continue to develop where they left off the previous year and produce seeds.

Tundra also occurs in mountain ranges above the tree-line, in altitudes too high for trees to grow, and in upper mountain meadows below permanently ice- and snow-covered summits. The conditions are similar to those of the arctic tundra, with high winds, low temperatures and impoverished soils frozen for much of the year. Many plants of the mountain tundra are identical to those of the Arctic. Plant geographers – botanists who study the distribution of plants – are trying to discover how these plants, which probably originated in the mountains, spread to the Arctic, often thousands of miles away. Some species may have made the trip as long ago as 10,000 years or so, following the receding ice sheets of the last Ice Age.

Many of the flowering plants of the tundra have brightly colored flowers that attract the few pollinating insects that live there. Other plants are wind-pollinating, such as certain catkin-bearing willows, or self-pollinating, as are some of the saxifrages. With some species the seeds germinate viviparously – before they separate from the plant – into plantlets ready to grow as soon as they touch the soil. In some other species, such as certain grasses, small plantlets develop from the flowers instead of from the seeds.

Deciduous Forests

In the ideal climatic conditions of the tropics, trees grow throughout the year. In the cold zones of the far north active tree growth, if not prevented by strong winds, occurs for only about two months each year. Between these extremes in the temperate regions trees grow for about half the year and rest for the other, less favorable half. During the cold season these deciduous trees, as they are called, have no leaves. They are thus better able to resist frost and excessive water loss.

Deciduous forests occur in a wide belt across Europe, eastern Asia and eastern North America and in smaller areas of the southern hemisphere. In prehistoric times deciduous forests covered almost all of these areas; only the interior of the continents were treeless grasslands. Since then, as the number of human beings increased and, with them, their requirements for food, shelter and fuel, about three-quarters of these forests have been destroyed. Europe and western Asia have been the most changed. The land where these trees once stood is now used for agriculture, housing, factories and recreation. Of the forests that remain, very little is unaffected by man. He has cut down the more useful trees, introduced tree species from one area to another, and brought his animals to graze on the forest floor. Many deciduous woodlands, however, are for these reasons now protected in national parks and nature reserves in an attempt to replenish the declining number of trees.

Woodlands are more complex than other plant

All the species of maple trees of the genus Acer, *have large leaves that turn brilliant scarlet in the fall creating an impressive display.*

Oak trees can live for hundreds of years. Some oaks in Britain are thought to be 900 years old. In a preserve of oaks, below, old specimens have become gnarled and moss-festooned.

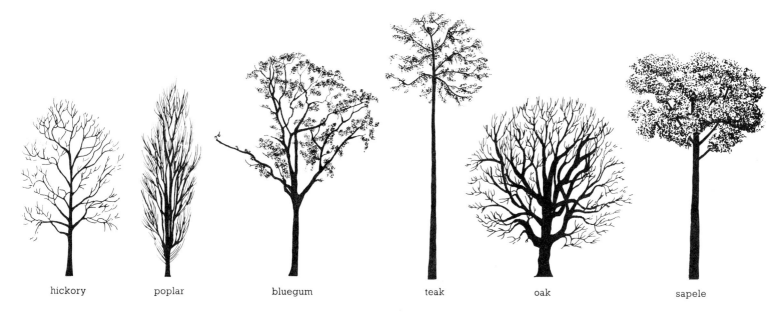

Trees can be recognized by their patterns of branching and overall shapes, above, as well as by their leaves, flowers and fruits.

Where some light can penetrate, as in a damp forest, ferns and young tree seedlings become a luxuriant vegetation.

communities, such as grasslands or deserts, because they consist of three or four layers of plants. The most important layer is the tree layer, composed of large, vigorous and mature trees. Underneath are the younger, smaller trees and shrubs. If the shade cast by the trees and shrubs is not too dense, under the shrubs will be a layer of early flowering herbaceous plants with their seedlings and the seedlings of shrubs and trees. Under the herbaceous layer, on the soil surface, there can be a covering of mosses, liverworts and lichens.

The appearance of a deciduous woodland changes markedly from one season to another as the trees lose their leaves and regain them several months later. This leaf-fall and regrowth affect the amounts of light and moisture that reach the shrub, herbaceous and ground layers, and so they have distinct and definite flowering and fruiting times. Thus in deciduous woodlands different plants can grow in different seasons, as the trees themselves change in appearance. Woodlands and forests, with their layers of plants and changing appearance, provide a wide variety of homes for animals, especially insects and birds. For this reason the deciduous woodlands of Europe and North America contain a great range of plants and animals and so are important in maintaining and controlling water vapor in the air and in preventing soil erosion by wind.

Conifer Forests and Gymnosperms

In Latin the word gymnosperm means naked seed. The gymnosperms are an important group of plants that form an evolutionary link between the spore-bearing ferns and the seed-bearing flowering plants. Although gymnosperms, like angiosperms, have seeds, the gymnosperms' seeds are not enclosed in an ovary as they develop. Instead, they are sheltered in a cone and, when ripe, are exposed on the scales of the cone for reproducing.

Gymnosperms first evolved about 350 million years ago. Among the early gymnosperms were the seed-ferns, or pteridosperms, and the conifers. The seed-ferns became extinct about 200 million years ago and are known today only through their fossilized remains, often found in coal. The conifers continued to evolve and gave rise to the modern conifers – spruces, yews, pines, redwoods, sequoias, cedars and cypresses. There are several less well-known gymnosperms, among them the maidenhair trees, or gingkos, the tropical cycads with palm-like leaves, and the South African *Welwitschia* plant.

Conifers are trees or shrubs and are found all over the world. They frequently form large tracts of forest, and as timber trees they are more widely planted than hardwoods, the flowering plant trees. There are relatively few species of conifers – about 550 – compared with more than 50,000 species of flowering plant trees.

Coniferous forests are found generally to the north of the zone of deciduous woodlands that lies across the temperate regions. There is nothing similar in the southern hemisphere because only the sea exists in the corresponding climatic zone. Canada and the Soviet Union have the largest areas of coniferous forest, but there are great tracts in northern Europe and across the northern United States.

Coniferous forest trees are generally less than 100 feet tall, but there are exceptions. In the redwood forests of California and in Oregon, the largest trees are more than 300 feet tall. Similarly, in the Olympic

Giant sequoias, Sequoiadendron giganteum, *are among the world's largest living examples of plants and animals. Most grow in national parks in the western United States, protected from both natural and man-made threats.*

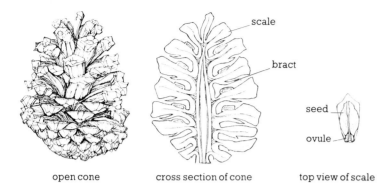

open cone · cross section of cone · top view of scale

A pine cone is hard and woody when ripe, left, and ready to release its seeds. Dry weather, or even fire, ripens cones, opening their scales and freeing the seeds within.

Cones can hang pendant-like, below left, as on this Douglas fir, Pseudotsuga menziesii, *a tree found chiefly in Washington and Oregon, or they may be carried erect, below, as in the Korean fir,* Abies koreana.

Peninsula of Washington, where rainfalls exceed eighty inches a year, the Douglas firs form a spectacular 200-foot-high forest with a luxuriant ground cover of small flowering plants, mosses and ferns. Rainfalls in most coniferous forests are low, about twenty inches a year, the growing season short – often only three months or less – and the soil is acid, infertile and frozen throughout the winter to seven feet below the surface.

A typical coniferous forest can be monotonous and gloomy. There are only a few dominant trees – pines, larches, spruces and firs. Where conditions are favorable, the conifers are sometimes intermingled with deciduous trees such as willows, alders, birches, aspens and poplars. The needle-like leaves of conifers only partly decay to produce an infertile soil in which little can grow and ground cover is further discouraged by the dense shade cast by the trees. Ferns, mosses, liverwort, lichens, herbaceous flowering plants, birds, insects and mammals all occur in coniferous forests, but with little diversity in any one place.

Despite the short growing season and generally poor conditions, conifers are well adapted to their habitat. Their leaves usually remain on the trees for at least three years before they are shed. Leaf fall takes place at all times but only a small percentage of the leaves is lost at any one time, and so the trees are always green. The leaf surfaces are thick and waxy, which ensures that water is not lost, and special chemicals in the cell sap prevent the sap liquid from freezing and thereby damaging the cells. The needle-shape of the leaves prevents snow from accumulating on them. Most coniferous forests are cold and windy, and therefore pollination of conifers is done by the wind.

Mountains and Islands

Although many species of plants grow in one or more adjacent countries, only a few species occur in nearly all regions. The common reed *Phragmites australis* and the bracken fern *Pteridium aquilinum* are found near the vicinity of the Arctic Circle to equatorial regions. Many other rare plants are confined to just one habitat or one small area of the earth's surface. Most plants that grow on islands and mountains fall into the latter category.

Islands separated from the mainlands for many thousands of years have developed species ideally suited to the peculiar conditions in which they live. Such species will seldom survive elsewhere in the wild and usually cannot cope with competition from other species introduced to the island, such as crops and weeds. Many of these islands endemics, as they are called, have become extinct, but unique collections still survive on such isolated islands as Socotra and the Aldabra Islands in the Indian Ocean, St. Helena in the South Atlantic, and the Hawaiian Islands in the Pacific.

For botanical purposes, mountains might be considered islands for the lowlands lying between them are just as much an isolating factor as the seas between islands. For this reason, mountain plants are often unique, and each peak in a high mountain range can have its own distinct set of species.

The zones of vegetation on a mountain are generally determined by altitude and slope, which affect wind speed, humidity, temperature, sunlight and soil stability. Trees are normally found on the lower slopes, and the summits can be covered by snow throughout the year; but in between there is an abundance of flowering plants, mosses, and lichens.

The peaks of the highest mountains are covered with snow and ice all year round, and the tops of many others are too cold and windswept for plant growth. But most other upland areas have plants and among them species of their own. The height at which upland plants are to be found varies greatly depending on such factors as the nearness of the sea, the distance from the Equator, the kind of rock and soil and the local wind pattern. The amount of direct sunlight can be important, too. In the northern hemisphere upland plants grow at a higher altitude on the sunlit, south side of a mountain than on the shaded and cooler north side.

Upland vegetation occurs under conditions that lowland plants cannot withstand. Mountain plants have to withstand harsher conditions than other plants and survive because they have become adapted to lower temperatures, higher winds and poorer soils. Many mountain species are covered with ice and snow for much of the year. They come into leaf, flower, produce their fruit and shed their seeds only in the short period after the winter snows have melted and before the first autumn frosts occur. Some plants are able to survive low temperatures even when in flower. Some specialized plants can be overtaken suddenly by winter conditions when in full flower. They then continue to complete their life-cycle six months later when the temperature rises again.

Mountain plants are often called Alpine-Arctics because they also grow in the Arctic Circle and in the Alps in Europe, as well as in Alpine climates all over the world. Rock garden plants are often mountain species. They do not always survive in gardens at low altitudes because the winter conditions are not cold enough for them to rest, and the soft growth they produce is not resistant to damp air and heavy rainfalls.

Although mountain plants are often picked and dug up, the absence of heavy grazing, mowing and plowing has enabled several species to survive. The grasslands and ledges, top, above the tree line in the Alps of Europe are often full of flowers.

The silversword plant shown at right is an endemic species, meaning that it grows only in a particular place, in this case Hawaii. This woody plant is made even more unusual by the fact that it evolved from a mainland herb.

Grasslands

The most common and most important plants in the world are the grasses. They provide food for most domesticated and wild animals, which in turn provide meat and milk. The grasses supply much of the world's sugar in the form of sugar cane. All the cereal crops, such as corn, wheat, oats, rye, barley, millet and rice, are grasses.

Grasses have fibrous roots. The growth buds either branch at ground level to form clumps or produce long stems underground or on the surface. The upright stems are usually hollow, tough and fibrous, and in bamboo they are quite woody. The leaves grow at more or less regular intervals on the stems, and their bases clasp the stem where a sheath protects the delicate growing tissues. Where the sheath meets the leaf blade there is a small, thin membrane, which is different in each type of grass and is useful for identifying the species.

Grass flowers, called spikelets, have no petals and are much smaller than the flowers of most other plants. The spikelets are usually grouped to form relatively large spikes, which in some plants may grow to as much as six feet long.

Every day millions of people around the world eat grasses in one form or another. Because they are easy to grow and produce crops quickly, they provide most of the world's supply of carbohydrates.

The seeds of all the cereals are packed with starch. The rest of the cereal plant, such as the stems and leaves, has food value too, as animal fodder and, after processing, for human consumption. Cereals can also be fermented to produce alcohol as an almost pollution-free fuel. As oil supplies are reduced, this use of cereals might become more important.

Natural grasslands occur where there is either too

The pictures below show some of the grasslands found around the world. In a savanna, left, the grass is short because of grazing animals and constant winds. Valleys, center, often have lush grasslands because they are protected from strong winds; water running off mountains or hills also keeps them moist. A typical woodland field, right, often contains a variety of grasses and wildflowers.

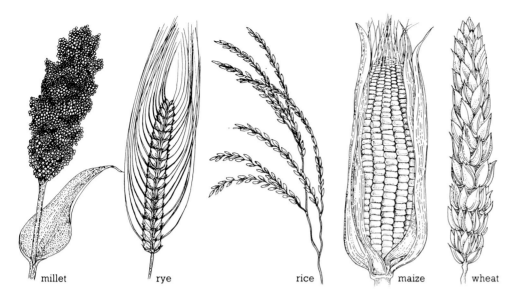

Cereal grasses such as those at left grow abundantly throughout the world. Millet, rye, and wheat prefer cooler climates than rice, which flourishes in hot, wet regions. Corn can thrive from Siberia to Argentina, providing it has high temperatures and adequate rainfall while it is maturing.

much wind or too dry or shallow a soil to support woodlands. They are thus found away from rain-bearing winds in the interior of continents, on the lower slopes of mountain ranges, or on hard rocks bearing poor soils. The world's largest areas of natural grasslands are the pampas of South America, the veldts of South Africa, the steppes of central Asia, the prairies of North America and large tracts in Australia and New Zealand. Where the climate is too dry or too hot for grasses and small herbaceous plants, deserts form.

Pioneers settling in uninhabited country cut down the trees and drain the soils. People have done this throughout Europe, North America, Asia and Australia. The forests have been replaced by grassland used for grazing cattle and sheep or plowed and sown for harvested crops.

The numbers and kinds of plants that grow in both natural and man-made grasslands vary according to the type of soil. Very acid soils, which are poor in minerals, support only a small number of different grasses. Chalk and limestone soils, which are both alkaline and rich in minerals, can carry a greater variety of species in a given area than any other plant community.

Deserts

These large saguaro cacti, which resemble candelabra, grow in Arizona and northern Mexico.

There are many kinds of deserts, all of them characterized by a lack of water available for use by plants. For this reason certain very salty or very cold areas, as well as dry and rocky ones are technically deserts.

Areas in which rainfall is very low – generally less than ten inches a year – and daytime temperatures are 80°F or higher, occupy almost a fifth of the earth's land surface. These areas are deserts, and some of them are very old, thanks to geographical factors that cause little rainfall and high temperatures. Many of the world's deserts are growing larger because of man. He cuts down forests and plows and overgrazes grasslands. Severe soil erosion follows and subsequent change in the climate of the region prevents native plants from re-establishing themselves. This process, known as desertification, is causing great concern to agriculturalists, especially in many of the developing nations.

However, deserts are not as empty of life as early explorers once thought. Plants that need little water and can tolerate extreme temperatures are widespread. Most of these plants are called succulents.

The most obvious feature of succulents is their water-filled leaves and stems. Some species, including most of the cacti, have lost their leaves and only the main stem remains. Many succulents live in areas of desert so hot and dry that their leaves and stems have few water-losing stomata. And even these pores are sunk in deep grooves that trap a layer of damp air. Water is retained also by certain chemicals within the cells. Succulents have extensive root systems. In a typical desert, plants stand far apart and below the surface the

Before flowering, these unusual Lithops salicola *from South Africa resemble the rocks on which they grow. They become conspicuous only when in flower.*

roots of each plant extend over a wide area, seeking water in the layers of soil and subsoil.

Succulents belong to many different plant families. Some large families have only a few succulent members, but in three almost every member is a succulent. The best known of these is the cactus family, Cactaceae, with 2000 species spread throughout the deserts of North, Central and South America. One genus, *Opuntia*, known as the prickly pear, grows in many parts of the world. When prickly pears were first introduced into Australia, they spread rapidly and became a threat to national food production. They were brought under control by the introduction of an insect that fed on them. Today, many cacti are popular garden and house plants although they require rather specialized conditions.

One subtropical family of succulents, the Aizoaceae, has large and often brilliantly colored daisy-like flowers, although some members of the family look like the pebbles and stones among which they live. The Crassulaceae family, with several species native to temperate Eurasia and North America, also has colorful flowers as well as succulent leaves in various shades of green, blue, and red, and often covered with a waxy coating like a ripe plum.

Most desert plants, including those below, have evolved an efficient system for coping with a lack of moisture and high daytime temperatures by storing water in their stems or leaves.

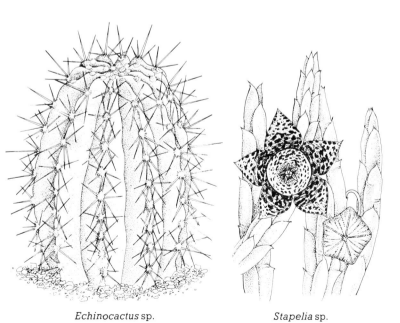
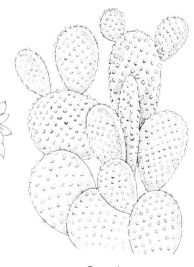

Echinocactus sp.   *Stapelia* sp.   *Cotyledon* sp.   *Opuntia* sp.

Tropics

The climate along the equatorial belt, except on the mountain ranges, is unlike that of any other place in the world, being uniformly hot and humid and with virtually no seasonal changes. Annual rainfall is usually eighty or more inches and daytime temperatures generally exceed 80°F. These conditions are ideal for plant growth and have given rise to equatorial, or tropical, rain forests.

Geological evidence suggests that these conditions have altered little since very early times. Even during the Ice Ages, when the tropics were cooler and drier than they are today, the climate favored plant life, and species that could not survive the freezing temperatures elsewhere took refuge there. The favorable conditions of the tropics also encouraged the evolution of new plant species and it has proceeded at a rapid rate. There are thirty times more species of plants in a tropical jungle than in a temperate deciduous woodland, and as many as 100 times more than in evergreen coniferous forests. In a tropical rain forest usually only a few specimens of the same species grow together in a single area.

Most tropical forest trees are broadleaf evergreen, straight, slender and 100 to 200 feet tall. Their bases are often supported by buttress or stilt roots. Flowering and fruiting periods often last for an entire year, with all the stages of development found on a single tree at any one time. These trees are well adapted to the excessively wet conditions of the rain forest. The leaves of most trees are shiny, have a marked central groove and often an elongated apex, or drip tip, so that the heavy rainfall and dew are channeled away and the leaves do not become too heavy and break off.

Trees of different species and various ages form several layers of dense canopy, under which only a few tree seedlings, ferns and parasitic and saprophytic plants – particularly fungi – can survive on the forest floor. In addition to trees and forest floor dwellers, many other plants grow in the tropical rain forest. Climbing plants, called lianas, often reach the tops of all but the tallest trees. Some of these climbers are so strong that they strangle and eventually kill the trees around which they are entwined.

The twigs, branches and sometimes the trunks and leaves of jungle trees can be covered with epiphytes – plants that grow on other plants for anchorage and not

Many of the choicest garden shrubs and trees in North America and Europe were introduced from subtropical, semievergreen forests such as this one on the slopes of the Himalayas where species of Rhododendron *and* Magnolia *may be seen in flower. Both are common garden plants.*

In this Venezuelan rain forest, a clearing has been created by a fallen giant tree. In this fertile climate, ferns, palms, and epiphytes abound, as do young trees.

as parasites. Mosses and lichens can grow epiphytically on other plants in all but the most heavily polluted and windiest sites. In the rain forests, orchids, ferns, bromeliads and many other kinds of plants live epiphytically, with their roots either dangling in the damp air or lying in fissures in the bark. The roots absorb minerals and small amounts of water from the decaying remains of leaves and other plant material on the bark. Some epiphytes collect water in cistern-like structures. Others, including many species of orchids, absorb water in spongy roots. More than 12,000 species of epiphytic orchids of tropical areas have been described. The areas just outside the tropical zones are the subtropics. The Everglades of Florida, for example, are mainly subtropical. Conditions here are similar to those of the tropics, except that rainfall may not be so high and there are seasonal changes. In the subtropics, epiphytes, lianas and dense areas of evergreen jungle occur, and animal life is abundant.

Both tropical and subtropical forests are threatened by the cutting of their trees and the subsequent use of the land for agricultural crops, road building, housing and industry. In both of these habitats, many species are becoming extinct before being properly classified.

Plants and Mankind

If it were not for the food manufacturing properties of green plants there would be no animals – including man. Man has always needed plants for food and also for medicine, fuel, shelter and clothing. In early times plants were made into weapons, too – bows and arrows, clubs and spears. Early man used the wild plants growing around him, but as civilizations developed many plants were cultivated in plots of land set aside for this purpose. Very likely agriculture as we know it today began when prehistoric man simply removed unwanted plants from among the ones he found useful.

Starting in earliest times, man took his useful plants with him on his migrations and used them as articles of trade. Today many of the world's major crop plants are not native to the regions in which they are grown. The wide range of timber, food, fiber and fuel plants now available is the result of careful selection of the more vigorous and productive plants in a crop and the crossbreeding of plants with useful qualities.

Weeds are usually defined as any plant growing where it is not required. A weed in one part of a garden can be a crop or decorative plant elsewhere. To gardeners and farmers, weeds are a nuisance because they compete with useful plants for water, minerals, light and space. If not controlled, weeds can overpower the cultivated plants and reduce their yield. Weeds may also have diseases and pests that spread to their cultivated neighbors.

Many weeds the farmer fights are foreign in origin and were first introduced with crop seeds from other countries. In their original countries they may have been kept under control by natural enemies, such as insects and diseases. But with no natural enemies present, the weeds grow vigorously. In the new environment they can be controlled by introducing appropriate insects and diseases but care must be taken that the crop plants are not attacked as well.

Many weeds are wildflowers from the surrounding countryside that find garden conditions more suitable than the wild where they have to compete with other plants. Herbaceous St. John's-wort, blackberries and dandelions are weeds that have spread to gardens and agricultural and forested areas all over the world. Willow herbs from Europe, North America and New Zealand are found as weeds in all of the cooler parts of the world.

Weeds are quick-growing plants that spread by producing large quantities of rapidly germinating seeds or have roots that creep outward to give rise to new plants in any available space. The weeds most

In the northern United States and in Canada rich crops of wheat are produced on fertile lands that were once prairies. The wheat is exported around the world.

Though highways scar the land and bring pollution, they are not entirely hostile to plant life. Wildflowers have colonized the banks of the highway above. In the ditch at the bottom of the bank, many species of water plants and insects flourish.

difficult to control are bindweed, quack grass, and similar plants. They all have long roots that break into small pieces when the soil is disturbed and each piece of root then develops into a new plant.

Many weeds appear only when growing conditions are ideal for cultivated plants. But several species flower, become fertilized, shed and germinate seeds at all seasons of the year and are unaffected by low temperatures. One example is the annual meadow grass, which occurs in many countries.

Plants are grown for decorative purposes by people everywhere. Usually the shape, scents and colors of the flowers or fruits are the attractive features. The textures, shapes and colors of the leaves and even the overall shape of the plant can also be prized.

Curative, preventative and stimulant drugs, as well as deadly poisons are produced by plants. These substances have led men to cultivate and worship these species and also to exploit and destroy them.

Of major concern today is that the past and continuing use of plants for all purposes is leading to their extinction. The need to protect species is generally accepted, and many voluntary and official conservation organizations now concern themselves with endangered species.

Plants as Medicines and Drugs

Throughout history plants have been used as tonics and remedies for illnesses. Modern research has questioned their powers, and many of the thousands of species of flowering plants, as well as a few ferns, conifers and some mosses, liverworts, seaweeds and fungi have been tested for their chemical contents and medicinal qualities. The results have shown that most of these plants have no curative value, but a few contain powerful and useful drugs. Although many of these drugs are now made artificially, several still come from plants cultivated to produce them more cheaply and efficiently than a factory can.

Quinine, prescribed for malaria, is extracted from the bark of the cinchona, a high-altitude tree found in the Andes and now grown in several parts of the world. The sedatives hyoscyamine, scopolamine and atropine, used to dilate the pupils of the eyes, are all derived from members of the nightshade family. Cocaine, which can be used as an anesthetic, is derived from the coca bush that grows on most continents. Eucalyptus oil comes from bluegum trees, many of which grow in Australia, and is used to treat head colds and ease breathing. Castor oil, a laxative, is made by pressing the seeds of the common castor oil plant. Morphine, a very powerful painkiller, is made from the sap of the opium poppies.

From earliest time people have used plants because of the sensations and moods they can induce. If some of the drugs extracted from plants – such as cocaine and morphine – are taken indiscriminately, they can become addictive. The user becomes dependent on them and requires increasing amounts, with the result that the drug can eventually poison and kill him.

A native of North America, the tobacco plant is cultivated throughout the world. The chemical nicotine, contained in dried and fermented tobacco leaves, is a

Papaver somniferum *is the poppy from which opium is extracted for medical and narcotic use. It is cultivated, often illegally, in many countries, where the milky sap from the seed pods is collected, dried, and processed to make morphine, codeine, and several other substances.*

This autumn crocus, Colchicum autumnale, *is cultivated as a decorative and drug-producing plant. The drug derived from this plant has been used to relieve gout for many centuries.*

The foxglove, Digitalis pupurea, *left, contains a drug widely used for the treatment of heart conditions. The deadly nightshade,* Atropa belladonna, *right, is an herb from which atropine is derived, a drug used to control muscle spasms and to dilate eyes for examination.*

drug with a mildly relaxing effect, but the tarry substances produced with it when chewed or smoked in a pipe, cigarette or cigar can be very harmful. Nicotine is also used for killing insects.

Tea, coffee and cocoa contain nonaddictive chemicals that are similar to, but milder than those in substances made from other drug plants. Cocoa contains the stimulant theobromine and is made from the beans of the cacao tree. The dried leaves of various evergreen shrubs are used to make tea, and the dried, roasted beans of coffee bushes are used for coffee. Both of these mild stimulants contain the drug caffeine.

Alcohol is a drug produced by fermenting sugars with the single-celled fungus yeast. Alcohol is used more than all the other drugs. A large part of the world production of alcohol is as an industrial solvent or fuel.

Insecticides are produced from several plants and are usually as effective as factory-made chemicals. Most have the advantage of not being poisonous to people, other mammals or birds and do not accumulate in the bodies of animals. One of the most widely used plants for this purpose is the pyrethrum.

Derris is made from the dried and powdered roots of the derris plant from Malaysia. As well as being used as an insecticide, it is used to kill fish, however, the fish can still be eaten immediately afterwards without harmful effects.

New Plants from Old

Plant cultivation began around early settlements, very likely when seeds from plants collected in the wild for food dropped onto the ground, germinated and grew into mature plants. By 9500 BC, in the Middle East, people had settled and had begun to grow crops rather than searching for all their food in the wild. About 2000 years later man not only grew many crops from seed but also selected and used the seeds from only the strongest, most productive plants.

This process of selection – using only the best plants – is still the most widespread method of producing new plants from old. Today, plants are selected not only for their yields but also for such features as resistance to disease, the ability to withstand bad weather and poor soil, and the speed and uniformity of maturing.

Occasionally a different individual unpredictably appears in a crop. Known as a mutation or a sport, it is the result of a spontaneous change. Many garden flowers are sports that have appeared among cultivated plants. Of the food plants, red and green cabbage, broccoli, cauliflower and Brussels sprouts all began as sports of the wild cabbage.

During the last sixty years new plants have been raised by artificial crossbreeding, or hybridization. The yield and quality of crop plants and the colors of the flowers of decorative plants have been increased by scientific breeding carried out by government research stations, universities and private seed firms and nurseries. Hybridization enables breeders to produce plants for particular conditions and needs. For example, potatoes that are resistant to disease, grow well in poor soil and are tasty as well have been produced through hybridization. Breeders start by pollinating two strains that each have desirable features. The hybrid may then be pollinated with one of its parents or a similar hybrid to reinforce certain desirable characteristics. This process is repeated until a worthwhile hybrid is produced.

This diagram shows how two corn plants that might be producing low yields can be carefully crossbred to create a plant that is larger, faster-growing, and more resistant to disease, frost and drought to provide the world with larger, more abundant produce.

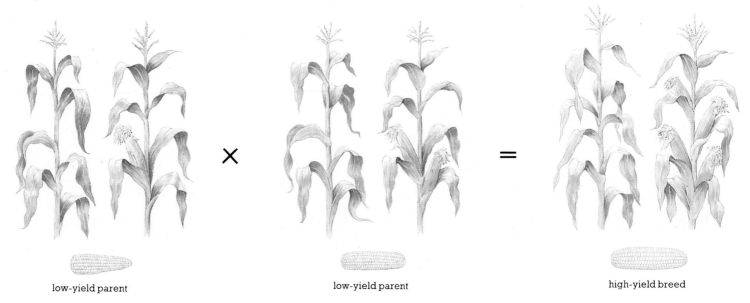

low-yield parent   low-yield parent   high-yield breed

The wild cabbage Brassica oleracea *grows as a cliff plant throughout Europe and the Mediterranean region. Over 1000 years of selection of the more palatable forms, and nearly 100 years of careful crossbreeding have produced hundreds of cultivars, or types, in several major groups. Thus today we have cabbage, top left, Brussels sprouts, top right, broccoli, bottom left, and cauliflowers, bottom right – all in a large range of races suitable for growing in different climates.*

In the past, breeding programs could take decades. Today they are greatly shortened, and a new plant can be created and disseminated in seven to fifteen years. The breeders speed up the process of hybridization by controlling such growing conditions as the amount of light and richness of the soil.

One early breeding program began in 1943 to develop high-yield wheat for Mexico and other Central American countries. The hybrids produced had the advantage of maturing quickly, so that in some areas two crops could be harvested in a single growing season. Later, soybeans and other crops were successfully hybridized and grown in various parts of the world. The so-called Green Revolution has increased food production in many countries.

But with the new high-yield strains have come new problems. Most of these plants need large quantities of chemical fertilizer to sustain the high yields. New pests and diseases have appeared, some of them particularly well-suited to prey on the new strains, many of which have not been successfully bred for disease resistance. Furthermore, the uniformity of the hybrid plants makes them uniformly vulnerable to the pests and diseases. In some cases, entire crops of some plants have been destroyed. These new crops must be protected until they are able to produce their own pest and disease-resistant characteristics.

Scientist are now selecting and breeding plants that could be used for food or fuel. Algae and fungi grown in tanks of diluted sewage and industrial waste may well be the food of the future.

Growing Plants

In many parts of the world, gardening is a major hobby. The selection of plants that can be grown in any garden depends on the climate – especially the temperature and rainfall – the kind of soil and how much care they require. Many garden plants need protection from wild and domesticated animals, birds, pests and diseases. It is also usually necessary to control the space between plants carefully so that they do not have to compete for water, light or air. Weeds have to be removed because they compete with garden plants and can carry pests and diseases.

To grow as wide a variety of plants as possible the soil must be kept rich and well watered. Nutrients such as manure, compost and chemical fertilizers must be added regularly and, if there is not enough rainfall, regular watering is essential. However, it is just as important to prevent the soil from becoming waterlogged by overwatering as it is to prevent it from drying out completely.

Many garden plants are not natives of the country in which they grow. Although some species will grow, flower and fruit in almost any area of the world, many garden plants will complete their life cycle from seed to seed only in their native areas. Otherwise many gardens would be overrun by newly introduced species, and many original plants would die out.

Plants can be grown outdoors in containers, such as pots and boxes. Some of the most successful plants to grow in pots are bulbs, such as daffodils, hyacinths and tulips. To prolong the flowering season as many as four layers of bulbs, separated by soil, can be planted in one pot. Those nearest the surface will flower first and those deepest will flower last.

Pots and boxes can also be used to grow climbers such as roses, ivies and annual fast-growing climbers such as nasturtiums and morning glories.

A garden can be expensive to stock if all the plants are bought from a nursery or garden center. In addition, the shock of transferring young plants from the luxury of a nursery to a new environment can stunt their

Even in a small backyard plot of thirty square yards enough vegetables and fruits can be grown to help feed a family. While many of the crops will be eaten as they are harvested, others, such as squashes, onions and carrots, can be stored, frozen, bottled or canned.

Indoor plants, left, are often grown for their attractive variegated leaves. Bonsai trees, such as the one shown above, miniature greenhouses and bottle gardens also make interesting and unusual houseplants especially when displayed together.

Bulbs and corms, such as the crocus corms below, will flower indoors in about eight weeks when planted in a bowl. These can make delightful and colorful houseplants.

growth or even kill them. However, nearly all garden plants can be grown from seeds, which are not expensive. As well, plants grown from seeds are better able to adjust to their environment from the time they germinate than plants bought in stores.

In some places it is not possible to have soil containers, and furthermore the soil itself, if it has to be bought, can be expensive. This problem can be overcome by putting stones, sand, even granulated or shredded plastic into a container and planting into it. These materials will anchor the plant, and food can be given by water with liquid manure. This soil-less method of growing is called hydroponics, and it can yield healthy, vigorous specimens. Although it is gaining popularity, very large plants are not as successfully grown hydroponically as small ones, and hydroponics is still an expensive way to grow crops.

Many people become specialist gardeners. Among the thousands of gardening clubs around the world are societies that cater to gardeners interested in a few species only. Thus there are clubs for growers of orchids, cacti and other succulents, bromeliads, ferns, roses, camellia and rhododendrons, trees, lilies, tulips and ivies. Such groups will meet once or twice a year to exchange information and advice and discuss the possibilities of creating new species.

Growing Trees

Trees are woody perennial plants that usually have a single trunk and can vary in height from miniature bonsai trees a few inches tall to the giants of the redwood forest that stretch 300 feet or more. Trees are conifers – cone-bearing – or flowering plants. Most of the conifers are evergreen, and only a few are deciduous and lose their needles at a set season. Of the flowering trees, most are deciduous.

Trees live much longer than most nonwoody plants. The trunk, branches and twigs usually increase in length each year, and many of the smaller, older and weaker stems are shed. Some stems stop growing in length after reaching a certain size but continue to increase in diameter. The outside of the stems of trees is protected by a bark composed of both living and dead cells. Some trees, such as cork oaks, are cultivated for their corky bark.

Sap is the tree's liquid contained in the veins, or sapwood. It can be extracted by making a slit in the trunk and piercing the veins. The sap from the rubber tree, which is called latex, is used to make rubber. Maple syrup is made from the sap of maple trees, particularly the sugar maple.

Trees are grown in almost every country and produce a great variety of crops. Most trees are grown as timber for building, furniture, fuel and paper manufacture. Trees are also widely grown for their fruits and seeds, to shelter agricultural land and even to attract birds and animals in nature reserves.

Almost everywhere that trees grow there are both native species and others that have been introduced. To provide trees for a greater range of habitats, and to improve features such as timber production, hardiness and disease-resistance, selected plants are crossbred, or hybridized for higher quality trees.

Decorative trees are grown in gardens, and parks and along sidewalks and highways. Many act as sound baffles and agents to filter out pollution. Trees are chosen for their shapes, the color of their leaves in the fall, and their bark, flowers and fruits. Before the leaves develop in spring, some trees produce flowers, such as the long, dangling catkins of alders, hazels and willows. The scents and colors of the opening leaf buds and young flowers attract pollinating insects as well as give pleasure to the passersby.

Successful tree cultivation demands careful siting of young trees. Single specimens rarely thrive in very windy or dry sites. Tall buildings and walls can create unsuitable, windy conditions, which can stunt the growth of trees or even kill them.

Trees need reasonably deep soil in order to anchor themselves and from which to take moisture during all

Wherever they are found, flowering trees are always a delight to the eyes. Here a blossoming camellia heralds the coming of spring.

While many thousands of trees are destroyed each year for use as timber, as many young seedlings are planted to ensure the continuation of the species. This orange grove in southern California will soon yield luscious oranges, to be exported all over the world.

Many trees are grown as a source of food. With careful crossbreeding special species of fruits can be cultivated that will be healthy, disease-resistant, and tasty. In this apple orchard, harvesting is being done by hand. In large orchards it is usually done by machines.

times of growth. Young trees should be supported by a stake. Trees are tolerant of most pests and diseases and most trees can withstand severe weather. In selecting a tree for planting it is essential to check the rate of growth, for, if the conditions in the garden are favorable, certain trees can grow so fast that other plants or even adjacent buildings can be destroyed by it. Some trees, such as poplars and cottonwoods can reach twenty-five feet in height and be expensive to remove.

It is possible to control the growth rate of trees by cutting out branches, but often the trees will retaliate by producing shoots, called suckers, in the soil around them which can overwhelm surrounding plants.

Trees can be grown from seeds or cuttings or by obtaining a plant from a nursery. Although it may take a few years, it is usually more satisfying and cheaper to plant a seed, which can be found under almost any mature tree.

Growing Shrubs and Climbers

Like trees, shrubs are woody perennial plants. Generally, however, shrubs do not have a single trunk, but many branches growing from the base, and do not grow as tall as trees. By careful pruning trees can be turned into shrubs and shrubs into trees. Even in the wild many trees can become shrub-like because of poor growing conditions, such as shallow soil and high winds. Shrubs occur throughout the world, either as undergrowth in woodlands or as a main feature of the landscape. Growing at the margins of woodlands, they extend the area of woody plant growth far into the adjacent grassland or desert.

Shrubs are widely grown in all countries, both as decorative garden plants and for crops. They provide many different crops, from fibers, such as cotton, and drugs, such as cocaine, to nuts and many soft fruits, such as raspberries and gooseberries. The wood of a few shrubs is used as timber for building or furniture. Several shrubs are grown for their long, straight stems and used as garden stakes and bean poles. In Europe, those of the hazel are used for making an ancient form of fencing called hurdles.

The Spanish, or sweet chestnut, is technically a tree, but periodically can be cut down to ground level to encourage the growth of many miniature trunks which are harvested every seven to ten years. Split into lengths they are used to make chestnut fencing, which is used in parks and gardens. Chestnut is a durable wood, and the fencing frequently lasts longer than the metal wire used to join it together.

Many flowering plants, both woody and nonwoody, have weak stems that require the support of another

Berberis darwinii

Hydrangea macrophylla 'Deutschland'

Skimmia japonica

Bryonia dioica, *the white bryony, left, is commonly found along roads and in hedges throughout Europe. The red berries of the plant are very poisonous, but few animals are tempted to eat them.*

Plants can climb by twining, but are usually helped by clinging hooks, left, suction disks, center, or clasping tendrils; right. Many use other plants for support and can eventually smother and kill them.

Four common garden shrubs, grown throughout North America and Europe are shown below. They either have only a very short main trunk, or have many stems arising at or just below soil level. Pruning keeps branches at the right size and shape.

Rhododendron 'Mrs P. D. Williams'

plant, rock or wall. Such stems climb, twine or scramble upwards and they can also creep along the ground. Some of these stems have clinging hooked thorns, clasping roots or tendrils. The force exerted by some twining plants, such as the tropical figs, can be so great that the tree on which the plant is climbing is strangled and eventually dies. Many climbing and scrambling plants are parasitic, obtaining all or some of their food from the living tissues of the host to which they cling.

Although they are too weak to stand upright without support, the stems of many climbers are very tough and have been used as string and rope by craftsmen in nonindustrialized countries for hundreds of years.

Climbing plants have long been favorites with gardeners because they can be used to cover unsightly objects such as dead tree trunks and garden sheds. Towns and cities are made much brighter by covering the fronts of houses and walls with vines such as Virginia creepers, whose large leaves turn brilliant red or orange in the fall.

Creeping plants, usually called ground covers, are used in gardens and orchards to cover otherwise bare ground and to reduce the growth of weeds. Creepers are planted on a larger scale on banks of loose soil and rocks to control soil erosion. Once embedded in the soil, the roots of these plants help keep particles of soil stable.

Herbaceous Plants

To the botanist a herbaceous plant is any flowering plant with no persistent woody parts. This category includes not only the ordinary garden and countryside plants, but also grasses, sedges, rushes, water plants, weeds and all bulbous and tuberous species. Herbaceous plants can be annuals, completing their life cycle from seed back to seed in a single year; biennials, which take two seasons; or perennials, which can last for several seasons.

To the gardener, herbaceous plants are those that die to the ground each winter and sprout to life or must be planted again when the growing season comes around. A conspicuous feature of many gardens is a long, narrow flowerbed filled with plants that provide a wealth of color throughout the spring and summer. A herbaceous border, as such flower beds are called, may in fact contain a few woody plants, but it is mainly used for growing a mass of closely packed herbaceous plants. There are hundreds of them, but among the most popular are marigolds and snapdragons.

In many gardens and parks, plants offering a wide range of flower colors are carefully chosen to provide a show throughout the spring and summer.

Orchids are the largest family of herbaceous plants. The lady's slipper, Cypripedium, *shown above grows in Malaysia and Thailand. Related species are found in Europe and North America.*

Most of the world's herbaceous plants originate in the tropics and cannot survive outdoors in Europe and North America, but many of them are ideal for growing in greenhouses. Among the favorite tropical herbaceous plants are orchids and bromeliads, the latter are members of the pineapple family and are watered through a cup formed by the leaves at the top of the plant. Several herbaceous plants, including the begonias and tropical vines and ferns, are cultivated not for their flowers but for their leaves.

The word herbaceous is often shortened to herb, and to most people herbs are plants, such as mint, thyme and sage, grown for the scents and flavors that their leaves, roots and flowers lend to food and drink. Certain herbs have long been grown for their medicinal value and are still often used instead of chemically produced drugs. Some herbs are used in the preparation of cosmetics and pesticides and as dyes.

One of the largest groups of herbs are the mints. There are many different flavors of mint and crossbreeding has produced many more. The most famous mints are peppermint and spearmint. Other mints

Although not all herbaceous plants are herbs in the culinary or medicinal sense, herbaceous borders of useful plants have been cultivated for nearly 1000 years in Europe. Many of these plants produce colorful and sweetly scented flowers as a bonus to their other qualities.

commonly grown for the flavors and scents of their leaves are sage, thyme and rosemary. The herbs fennel and dill are grown for their flavored leaves and seeds, and ginger and licorice for their roots.

The cereals, most of the root crops and the salad plants are technically herbaceous. The many varieties of herbaceous *Brassica oleracea* include cabbage, kale, Brussels sprouts, cauliflower, broccoli and kohlrabi, all grown in cool, temperate regions.

Orchids are found in nearly every country and type of countryside, from the tundra of Greenland to the hot, humid rain forests of Borneo and New Guinea. Orchids are grown for their colorful, attractive appearance. Their flowers are unusual in that one of the three petals is a noticeable lip that attracts pollinating insects and can be a landing platform for them.

Garden chrysanthemums, members of the daisy family, are grown all over cooler parts of the world and there are native species in many countries. Many of the large, mop-headed flowers seen in shops and in nurseries are hybrids, the first of which came from China and Japan hundreds of years ago. The flowers on these are really masses of little flowers, or florets, grouped closely together.

Conservation and Threatened Species

Since plants first evolved, many species have been unable to adapt to changing conditions and thousands of them have become extinct. But the extinction rate of plants has never been as fast as the rate of evolution of new species, and so there is a greater diversity of plants in the world today than ever before. Still, there is great concern today over the near-extinction of certain plants, particularly those from the tropical rain forests.

All living things – both animals and plants – live in what is called an ecosystem, which includes the plants and animals themselves, water, air, soil and rocks, and the relationships among all of them. Ideally, all the parts of an ecosystem are delicately balanced, with no one part overly affecting the others. Obviously, a sudden flood, a volcanic eruption, or an earthquake can temporarily upset the balance, but it is usually soon restored. A new species can also disrupt an ecosystem, but before long it either contributes towards the balance, or becomes extinct. Only one species that has appeared on earth persistently upsets the balance of any ecosystem it enters and seriously reduces an ecosystem's natural diversity – and that species is man.

For the first two million or so years that people lived on earth they were part of the ecosystem, but about 11,000 years ago they started to live in settled communities growing crops and domesticating animals which grazed on land that was originally covered by natural vegetation. When the food supply was no longer a major factor controlling human reproduction and survival, the species became more abundant and destroyed more plants to provide space for its own activities. In 1975 there were more than four billion human beings: By 1999 there could be six billion and their effect on the environment and the plants of the world could be disastrous.

Human activities that cause the destruction of plants include the draining of marshes, the removal of trees and hedges and the soil erosion that ensues, peat removal, excessive use of weedkillers, pollution and the introduction of artificially bred plants that overwhelm natural species. But as mankind's destructive activities have increased, so has his awareness of them.

Plants are conserved in different ways, depending on their life cycles, their habitats and the threats they face. Ideally a plant is conserved in its natural habitat the wild, but if its habitat is damaged or destroyed the plant must be removed to a garden or greenhouse to grow and reproduce.

Many rare plants are conserved in the wild in protected nature reserves. There, all threats to them are controlled so that plants may produce seeds and multiply. As a safeguard against some very rare plants becoming extinct, their seeds are collected and stored in seed banks. Under carefully controlled humidity, temperature and pressure, most plant seeds retain their ability to germinate for many years, much longer than if they were kept in ordinary conditions.

Scientists are studying other environments in which plants can grow. As part of the space exploration programs the effects of weightlessness on certain plants is studied. These plants turn out to be much more responsive to changes in the quantities and qualities of their food than to speed or gravity. Experiments with flowering plants have also shown that the light, temperature and composition of the air around them are still the most important factors, just as they are on earth.

This photograph shows the luxuriant and diverse nature of an upland rain forest in northern Queensland, Australia. Tropical rain forests are being destroyed throughout equatorial regions with the result that many hundreds of species are becoming extinct. In rapidly developing countries such as Australia, the forests could become extinct in the next fifty years if action is not taken now.

Glossary

Algae: (singular, *alga*) a group of flowerless plants, some single-celled and a few are able to swim

Angiosperms: flowering plants

Anther: the part of the stamen, generally at the tip, that produces the pollen

Asexual reproduction: reproduction not involving the fusion of male and female cells

Axillary buds: lateral bud situated at the junction of a leaf-stalk and stem

Bacteriophage: a virus that attacks single bacterial cells and causes them to produce more bacteriophages

Bulb: an underground part of a plant that stores food in its leaves so the plant can survive an unfavorable season. Also reproduces the plant in the growing season

Cambium: layer of active cells each of which divide into a new cambium cell to create such tissues as xylem, phloem and cork

Cellulose: a carbohydrate that strengthens cell walls and forms fibers in stems

Chlorophyll: a green coloring matter that is found in all green plants; needed for photosynthesis

Chloroplast: solid body in a plant cell that contains chlorophyll; the site of photosynthesis

Chromosome: thread-like mass of genetic material inside the nucleus of a plant cell

Collenchyma: plant tissue of living cells with thickened cellulose walls; forms strengthening tissue

Corm: a short, underground stem that stores food and reproduces the plant

Cuticle: outer protective layer of a leaf made of cutin secreted by epidermis cells

Cytoplasm: the active, living material of a cell, excluding the nucleus, vacuole and cell wall

Dicotyledon: the larger of the two classes of the angiosperms, characterized by having two seed leaves, or cotyledons

Endosperm: food-reserve tissues in seeds

Epidermis: outer layers of cells of most parts of a plant

Epiphyte: a plant growing on another plant for support and not taking any food from it

Evergreen: a plant that keeps its leaves for more than a year, not shedding them at any particular time

Fern: a flowerless plant that has leaves with veins

Fungus: (plural *fungi*) a simple plant that cannot make its own food but takes it from dead or living plants or animals

Hybrid: the result of crossbreeding two plants that are not alike

Hyphae: the individual thread-like part of the body of a fungus

Inflorescence: a group of flowers on a single main stem

Lichen: a plant that consists of an alga and fungus growing together for mutual benefit

Liverwort: a flowerless plant without veins and with a life cycle in which a spore-producing phase alternates with a dominant gamete-producing phase

Meristem: the region of actively dividing cells found at the tip of a stem or root

Monocarpic: flowering once and then dying

Monocotyledon: the smaller of two classes of the angiosperms, characterized by having a single seed-leaf or cotyledon

Mycorrhiza: the association, usually for mutural benefit, between a fungus and the roots of a plant

Node: the region of a stem where a leaf or secondary branch grows

Nucleus: the control center of a cell, containing the genetic material derived from previous generations

Ovary: in flowering plants, the part of the female reproductive system in which the ovules are found

Ovule: female sex cell found in the ovary. After fertilization by a male sex cell each ovule turns into a seed

Parasite: a plant that lives on or in another living plant or animal, taking food from the host organism without giving any in return

Parenchyma: nonspecialized cells making up the bulk of plant tissue

Petiole: stalk of a leaf

Pistil: female organ of flowers made up of an ovary, style and stigma

Phloem: part of the vein of a plant used for transporting foods

Photosynthesis: the process by which plants convert energy from the sun with the aid of chlorophyll, oxygen and water to produce sugars

Pollen: the powder-like mass of minute spores that, when germinated, produce the male sex cells of a flowering plant

Protoplasm: living contents of a cell

Respiration: the process by which a plant converts stored foods into energy

Rhizome: an underground creeping stem, usually able to store food

Saprophyte: a plant that lives in or on a dead plant or animal and takes food from it

Sclerenchyma: plant tissue of living or dead cells with walls impregnated with lignin; forms strengthening tissue

Spore: an asexual reproductive body that germinates to produce sex cells or a new plant

Stamen: male organ of flowers made up of a stalk and anther containing pollen grains

Stigma: part of the female stalk of a flower that catches pollen

Stipule: small, usually leaf-like projection at the base of a leaf-stalk

Stoma: (plural *stomata*) a pore on a leaf that opens and closes to allow air and water vapor to pass in and out of the leaf

Terminal bud: bud at the top of a shoot from which main growth of plant continues

Testa: the seed coat

Thallus: plant body not divided into separate stem, leaves and root; generally applied to the body of seaweeds and some toadstools

Transpiration: the flow of water from the root of a plant, through the stems into the leaves and out into the air

Tuber: an underground stem or root swollen with food reserves

Vacuole: part of a cell containing the sap; a mixture of water and dissolved materials

Xylem: water-conducting tissue of plants

Index

algae 8, 23, 35, 44-7
   origin 7, 36-7
   reproduction 26
alpine habitats 64-5, 71
angiosperms 16, 27, 36-7
annuals 33
aquatic plants 62-3
asexual reproduction 26-7

bacteria 7, 35, 40-1
bacteriophages 43
bark 13, 86
behaviour responses 34-5
bonsai trees 85, 86
bracken fern 57, 70
breeding new plants 82-3
buds 12-13, 16, 21
bulbs 20-1, 85

cacti 74-5
cambium 13
cells 8-9, 12, 32-3
   algae 44, 46
   bacteria 40
cereal grasses 72-3, 78
chlorophyll 8, 22, 30, 33
classification 38-9
   algae 44-7
   ferns 56
   fungi 49
   palms 58
climatic zones 60-1
cones 16, 69
conifers 68-9
conservation 79, 92-3
corms 20-1, 85
cotyledons 31
cyanobacteria (Cyanophyta) 7, 41
cytoplasm 8-9

deciduous forests 66-7
desert habitats 74-5
diatoms 44-5
dicotyledons 31
dinoflagellates 46
disease organisms 40-1, 42, 49
DNA 9, 33, 42
drugs from plants 80-1

ecosystems 92
embryos in seeds 18, 30
epiphytes 10, 76-7
evolution 6-7, 36-7

female organs, flowers 9, 16-17
ferns 7, 14, 56-7

origins 36-7
   reproduction 26-7
flowering plants 7, 8-21
   origins 36-7
   reproduction 26-7
flowers 9, 16-17, 28-9, 33
food plants 82-4, 87, 91
forests 66-9, 76-7, 92-3
fossils 6-7, 37, 38
   algae 44, 45
   ferns 56
fruit 18-19
fungi 11, 23, 48-51
   movement response 35
   origins 7, 37

garden plants 84-5, 89-91
germination 30-1
grasses 10, 14
grassland habitats 71, 72-3, 78
growing methods 84-91
growth 30-5
gymnosperms 16, 27, 68-9

habitats 60-77
herbaceous plants 90-1
Hooker, Dr Joseph Dalton 39
hormones 33, 34-5
horsetails 7, 37, 56-7
houseplants 59, 85
hybrid plants 82-3

insecticide plants 81
insectivorous plants 34-5
island habitats 70-1

leaves 9, 14-15
lichens 52-3, 65
life-cycles 33, 36-7
Linnaeus, Carolus 38, 39
liverworts 7, 26, 37, 54-5

male organs, flowers 16
mankind's effect 76-7, 92
marine algae 44-7
medicinal plants 80-1
monocotyledons 31, 58
mosses 7, 26, 37, 54-5
mountain habitats 64-5, 70-1
mushrooms 23, 30-1
mycorrhizal fungi 11

naming plants 38-9
nectar secretion 17

orchids 11, 77, 90, 91
origin of plants 6-7, 36-7

palms 58-9
penicillins 48-9
perennials 33
petals 9, 16-17

petioles 14
phages 42-3
phloem 9, 37
photosynthesis 8, 22-3, 30, 44
   leaves 14, 15
phytoplankton 44-5, 46
pollination 16-17, 28-9

reproduction 26-9
   flowering plants 9, 16-19
   fungi 48-9, 51
   gymnosperms 68, 69
   lichens and mosses 53, 55
   seaweeds 46-7
respiration 24
rhizomes 13
roots 10-11, 30-1, 41
   desert plants 74-5
   water transport 24-5

savanna (grassland) 72
seashore plants 63
seaweeds 46-7
seeds 18-19, 30-1, 32
sepals 16-17
sexual reproduction 26-7
shrubs 67, 88-9
spores, fungi 48-51
stems 12-13
stipules 14
stomata 14, 22, 24-5
storage organs 10, 20-1
structure of plants 8-21
subtropical areas 77
succulents 74-5
symbiosis 11, 52

testa 18, 30
toadstools 50
transpiration 24-5
traumatin 34-5
trees 10, 11, 13, 76, 66-9
   cultivation 86-7
tropical rain forest 76-7, 92-3
tropisms 34-5
tubers 8, 20-1
tundra habitats 64-5

vegetation zones 60-1
veins (vascular tissue)
   ferns and mosses 55, 56
   plants 9, 10, 12, 14-15
viruses 42-3

water transport 10-11, 12, 24-5
weeds 78-9
wetlands habitats 62-3
woodland 60, 66-9

xylem 9, 12, 33

yeasts 48, 81

Credits

The Publishers gratefully acknowledge permission to reproduce the following illustrations: A-Z Botanical 17*b*, 44*br*; Heather Angel 19*t*, 26*t*, 29*tr*, 34*bl, br*, 35*tl, tr*, 48*t*, 59*r*, 69*r*; Ardea 28, 38, 45, 46, 53*br*, 54, 62*t, br*, 53*l*, 63*r*, 70, 76, 79*t*, 81*t*, 93; Dr. Lloyd M. Beidler/Science Photo Library 8, 24*l*, 29*tl*; C M Clay/National Vegetable Research Station 43; Bruce Coleman Ltd., 6, 7*l, r*, 19*bl*, 20, 23*tr, b*, 31, 34*t*, 41, 44*tl*, 49, 51*b*, 52, 53*t*, 55*t, b*, 57*r, b*, 51, 64*t, b*, 68, 71*t, b*73*l, r*, 74, 90*b, t*, 91; Gene Cox 13*l, r*, 14*t, b*, 30, 40, 44*tr*, 48*b*; Gene Cox/University of Bath 40*b*; Daily Telegraph Colour Library 80; Aubrey Dewar 3, 17*tl, tr*; Eastman Kodak Company & Kodak Ltd. 16; Dr. Patrick Echlin/University of Cambridge 29*bl, br*, Mary Evans Picture Library 39, R & C Foord 23; Brian Furner 21, 83*tl, tr, bl, br*, 85*tl*; N. Gryspeerdt 85*tr*; R. Harding 51*t*; R. D. Hunt 19*br*, 81*bl*; Baron Hugo Van Lawick 72; Gordon Leedale/Biophoto Associates 44*bl*; Massey Ferguson 79*b*; Nature Photographers Ltd. 35*bl, br*, 53*bl*, 57*tl*, 66*l*, 75; Oxford Scientific Films 11*t*, 56, 58, 59*l*, 65, 66*r*, 67, 69*l*, 77; Science Photo Library 9*ll*, 9*tc*, o*lr*, 24*r*; John Topham Picture Library 11*b*, 12, 23*tl*, 60, 84; Trewin Copplestone Publishing 85*b*; Vision International 86; ZEFA 87*t, b*; H H Heunert/Zeiss Inforation 26*B*.

Artwork by: Linda Broad 47, 73, 75; Carol Kane 67; Carol McCleeve 33, 42, 43, 61; David & Theo Nockels 16, 18, 69; Richard Phipps 12, 15, 32, 36-37, 82; Paddy Sellars 14-15, 38, 50; Charlotte Styles 8, 10, 21, 22, 25.

Cover photograph: Bruce Coleman Limited

Index by: Indexing Specialists, Hove

Bibliography

Pictorial Dictionary of the Plant World, M. Chinery, Sampson, Low, Marston, 1967
Anatomy and Activities of Plants, C. J. Clegg and Gene Cox, John Murray, 1978
Flowering Plants of the World, V. H. Heywood (Ed), Oxford University Press, 1978
Botany, Terry L. Hufford, Harper and Row, 1978
Plants, Victor A. Greulach and J. Edison Adams, John Wiley, 1967
Biological Science, William T. Keeton, W. W. Norton, 1980
The Plants, Frits W. Went, Time-Life, 1971
Flowers; Trees; Orchids; Non-Flowering Plants, (Golden Nature Guides), Herbert Zim et al, Golden Press, 1978